70th Annual Meeting of the Labor and Employment Relations Association and LERA@ASSA Meeting 2018

Labor and Employment Relations Association Series

Philadelphia, Pennsylvania, USA
4-7 January 2018

Baltimore, Maryland, USA
14-17 June 2018

Editors:

Ariel Avgar

ISBN: 978-1-5108-9509-6

Printed from e-media with permission by:

Curran Associates, Inc.
57 Morehouse Lane
Red Hook, NY 12571

Some format issues inherent in the e-media version may also appear in this print version.

Printed with permission by Curran Associates, Inc. (2019)

For permission requests, please contact Labor and Employment Relations Association (LERA) at the address below.

Labor and Employment Relations Association (LERA)
University of Illinois at Urbana-Champaign
121 Labor and Employment Relations Bldg.,
504 E. Armory Ave.
Champaign IL 61820
USA

Phone: (217) 333-0072
Fax: (217) 265-5130

LERAoffice@illinois.edu

Additional copies of this publication are available from:

Curran Associates, Inc.
57 Morehouse Lane
Red Hook, NY 12571 USA
Phone: 845-758-0400
Fax: 845-758-2633
Email: curran@proceedings.com
Web: www.proceedings.com

LABOR AND EMPLOYMENT RELATIONS

ASSOCIATION SERIES

Proceedings of the LERA 2018 Meetings

LERA@ASSA Meeting
January 4–7, 2018, Philadelphia, PA
(in conjunction with ASSA/AEA)

and

LERA 70th Annual Meeting,
June 14–17, 2018, Baltimore, MD

Ariel Avgar, Editor-in-Chief

First Edition

LABOR AND EMPLOYMENT RELATIONS ASSOCIATION SERIES
(Formerly the Industrial Relations Research Association)
Proceedings of the Annual Meeting
Annual Research Volume
Perspectives on Work Magazine

Inquiries and other communications regarding membership, meetings, publications, and general affairs of the Association, as well as notice of address changes, should be addressed to the LERA office.

LABOR AND EMPLOYMENT RELATIONS ASSOCIATION
University of Illinois at Urbana-Champaign School of Labor and Employment Relations
121 Labor and Employment Relations Building, MC-504
504 East Armory Avenue, Champaign, IL 61820
Telephone: 217/333-0072 Fax: 217/265-5130 E-mail: LERAoffice@illinois.edu
Website: http://lera.memberclicks.net EPRN: http://www.employmentpolicy.org/

CONTENTS

LERA 70TH ANNUAL MEETING, JUNE 14–17, 2018, BALTIMORE, MD

LERA@ASSA MEETING, JANUARY 4–7, 2018, PHILADELPHIA, PA

VI.
INCOME, WORK, AND WEALTH: UBI, JOB GUARANTEES, BABY BONDS, AND CARE WORK

Teresa Ghilarducci, Presiding

VII.
SKILL MATCHMAKING IN THE MODERN ECONOMY

Susan J. Schurman, Presiding

VIII.
LERA ANNUAL REPORTS

LABOR AND EMPLOYMENT RELATIONS ASSOCIATION SERIES

Proceedings of the LERA 2018 Meetings

LERA@ASSA Meeting
January 5–7, 2018
Philadelphia, Pennsylvania

LERA 70th[th] Annual Meeting
June 14–17, 2018
Baltimore, Maryland

OFFICERS OF LERA FOR 2017–2018

PRESIDENT
Harry C. Katz, Cornell University

PRESIDENT ELECT
Kris Rondeau, AFSCME

PAST PRESIDENT
Janice Bellace, University of Pennsylvania

SECRETARY-TREASURER
Craig Olson, University of Illinois at Urbana-Champaign

EDITOR-IN-CHIEF
Ariel C. Avgar, Cornell University

CHAPTER ADVISORY CHAIR
William Canak, Middle Tennessee State University

EXECUTIVE DIRECTOR
Emily Smith, University of Illinois at Urbana-Champaign

EXECUTIVE BOARD MEMBERS
Sylvia Allegretto, Ctr. on Wage & Empl. Dynamics
Annette Bernhardt, Univ. of California, Berkeley
Matthew Bodah, Univ. of Rhode Island
John Budd, Univ. of Minnesota
Bill Dirksen, Ford Motor Company
Virginia Doellgast, Cornell Univ.
Jonathan Donehower, Kaiser Permanente
Eric Fidoten, Campbell Soup
Michele Hoyman, Univ. of N. Carolina at Chapel Hill
Joan Husted, Hawaii State Teachers Association (ret.)
Charles Jeszeck, Government Accounting Office
Patrice Mareschal, Rutgers Univ.
Brad Markell, AFL-CIO
Daniel Marschall, Working for America Institute
Sheila Mayberry, Arbitrator and Mediator
Jim Pruitt, Kaiser Permanente
Saul Rubenstein, Rutgers Univ.
Jeffrey Wheeler, Georgetown Univ.

LEGAL COUNSEL
Steven B. Rynecki, von Briesen, Purtell & Roper, S.C.

PAST PRESIDENTS OF LERA

1948—Edwin E. Witte
1949—Sumner H. Slichter
1950—George W. Taylor
1951—William Leiserson
1952—J. Douglas Brown
1953—Ewan Clague
1954—Clark Kerr
1955—Lloyd G. Reynolds
1956—Richard A. Lester
1957—Dale Yoder
1958—E. Wight Bakke
1959—William Haber
1960—John T. Dunlop
1961—Philip Taft
1962—Charles A. Myers
1963—William F. Whyte
1964—Solomon Barkin
1965—Edwin Young
1966—Arthur M. Ross
1967—Neil Chamberlain
1968—George P. Shultz
1969—Frederick Harbison
1970—Douglass V. Brown
1971—George Hildebrand
1972—Benjamin Aaron
1973—Douglas H. Soutar
1974—Nathaniel Goldfinger
1975—Gerald G. Somers
1976—Irving Bernstein
1977—F. Ray Marshall
1978—Charles C. Killingsworth
1979—Jerome M. Rosow
1980—Jack Barbash
1981—Rudolph A. Oswald
1982—Milton Derber
1983—Jack Stieber
1984—Wayne L. Horvitz
1985—Everett Kassalow
1986—Lloyd Ulman
1987—Michael H. Moskow
1988—Phyllis A. Wallace
1989—Joyce D. Miller
1990—Robert B. McKersie
1991—James L. Stern
1992—Ernest J. Savoie
1993—George Strauss
1994—Lynn R. Williams
1995—Walter Gershenfeld
1996—Hoyt N. Wheeler
1997—Francine D. Blau
1998—F. Donal O'Brien
1999—Thomas A. Kochan
2000—Sheldon Friedman
2001—Magdalena Jacobsen
2002—John F. Burton, Jr.
2003—Paula B. Voos
2004—Marlene Heyser
2005—Stephen R. Sleigh
2006—David B. Lipsky
2007—Eileen B. Hoffman
2008—Anthony Oliver, Jr.
2009—Joel Cutcher-Gershenfeld
2010—Eileen Appelbaum
2011—Gordon Pavy
2012/13—David Lewin
2013/14—Lisa Lynch
2014/15—Martin Mulloy
2015/16—Bonnie Castrey
2016/17—Janice Bellace

I. 2018 Presidential Address

The Expansion and Implications of Various Forms of Collective Representation in the United States

HARRY C. KATZ
Cornell University School of Industrial and Labor Relations

Introduction

Before my formal remarks, I would like to comment on the state of LERA. I am pleased to report that LERA is healthy—we experienced strong growth in our membership over the last several years, including a noteworthy increase in the number of student members. Attendance at this meeting is at a record level. Due to those things and the intelligent hard work of our University of Illinois–based staff and the helpful contribution that many of you make as volunteers, LERA is now on a solid financial footing. Yet we should not be complacent. I encourage LERA to continue to strive to be inclusive in its programming and to seek the active involvement of the diverse parties involved in the world of work.

In my remarks, to which I now turn, I will describe how changes in the nature of collective action in the United States are leading to a broad array of groups who provide collective representation to employees but do so without making use of traditional collective bargaining agreements. The spread and increasing importance of those diverse forms of collective representation are among the many reasons that, to stay relevant, LERA must become ever more inclusive.

The Growth of Collective Representation

I have been a student of collective bargaining my entire career, as much of my scholarship and teaching has focused on understanding why and how U.S. collective bargaining evolved over the post–World War II period. What I am now struck by is that various new organizations are being used by employees to pursue group action so as to improve those employees' terms and conditions of work. Let me first describe how I came to see this emerging trend, as the origins of my thinking lead me to a related point about this development—namely, that the U.S. labor relations system is becoming increasingly similar to the labor relations systems that exist in emerging countries.

With Tom Kochan and Alex Colvin, three years ago I published a textbook, *Labor Relations in a Globalizing World* (Katz, Kochan, and Colvin 2015). In that book, we trace how core principles about bargaining power and negotiations can be used, and in some cases appropriately modified, to describe labor relations in emerging countries. We focus in particular on recent developments in China, India, Brazil, and South Africa as case illustrations and also focus on those countries because they are major players in the global economy. As we describe, in emerging countries the activities of nongovernmental organizations (NGOs) have become a more significant influence than trade unions on employees' terms and conditions of employment.

This year, Tom, Alex, and I published the fifth edition of our U.S. collective bargaining textbook (Katz, Kochan, and Colvin 2017.) In that book we spend a significant amount of space describing how NGOs have become a significant force within U.S. labor relations. Perhaps it was the fact that we had been alerted to the role NGOs are playing in emerging countries that led us to recognize the influence of NGOs in the United States as well, as the growing influence of NGOs was becoming the subject of labor relations research and current events.

Author's address: Cornell University, 356 ILR Research Building, Ithaca, NY 14853

These NGOs include worker rights groups such as the Immokalee Coalition (Brudney 2016; Marquis 2017). They also include the worker and immigrant rights centers and organizations that Janice Fine has brought to all of our attention (Fine 2006). The affinity groups that have emerged within many non-union companies represent another type of NGO focused on improving work conditions, where the inadequacy of existing complaint procedures at Uber and other companies has become a key issue (Maffie 2018a, 2018b). The recent grassroots protests of teachers in West Virginia, Oklahoma, and Arizona can be viewed as another type of collective representation because they involve mass action, focus on improving teacher pay and work conditions, and don't involve formal collective bargaining.

The various collective representation campaigns just mentioned do not involve unions, yet it is interesting to note that even where unions are present within normal collective bargaining, including cases in which unions have a significant amount of bargaining power, those unions have become involved in issues that had not been addressed in collective bargaining agreements and were not part of standard union demands. Union involvement in work restructuring—including joint steering committees, team work systems, and other forms of joint governance—which began in the early 1980s, should be recognized as an example of this type of collective representation, which broadens the nature of collective action.

I have been studying several collective bargaining relationships where unions recently have become extensively involved in the determination and administration of employee health benefits. These are the voluntary employee beneficiary associations (VEBAs)—set up to run the UAW–represented retired workers at the Big Three auto companies and the systems providing health benefits to New York City's 340,000 unionized public employees. In both of these cases, union involvement in the determination and administration of health benefits goes far beyond the traditional influence unions exerted in past bargaining. I see this involvement as an illustration of how collective representation can supplement and broaden collective bargaining. I find it particularly interesting that these latter examples of growing collective representation are taking place where unions are strong and are not leading to a diminished role for those unions.

Criteria That Can Help Inform the Assessment of Collective Representation

To help ascertain the influence and staying power of these various forms of increasing collective representation, the following criteria should be assessed:

- The degree to which employees exert an independent influence and voice via this representation—in other words, the degree to which a given representation process is free of managerial dominance.
- The breadth and depth of the issues addressed by a representation process. Efforts that focus on a single issue such as the Fight for $15 should be seen as advocacy rather than representation. (Note: this is not to diminish the potential importance of this and other advocacy campaigns.)
- The extent to which leaders of a representation effort are democratically elected. This relates closely to the first criterion.
- Perhaps most influential on the staying power of any representation process, whether that process is being institutionalized through agreements or procedures that not only clarify the channels of employee voice but also provide a sustainable financial basis for a representation process.

In addition to these criteria, a central question for any and all collective representation processes is the degree to which they serve as a complement or substitute for traditional collective bargaining. The labor movement is rightly worried that many of the current representation initiatives might one way or another dissuade employees from seeking or supporting union representation. Where management dominates a given employee representation process, it is reasonable to suspect that a key motivation for management in those settings is to reduce employee interest in unionization. While this is a legitimate worry, it is instructive to remember research showing that, historically, many employee associations that initially shunned collective bargaining, such as the National Educational Association and police and firefighter benevolent associations, did provide meaningful collective representation and helped stimulate and lay the groundwork for the eventual wave of public sector collective bargaining that emerged in the early 1960s (Freeman and Ichniowski

1988). The lesson from the public sector is that collective representation can serve as a stimulant to collective bargaining and not necessarily be a substitute for such bargaining.

Diversity in Employee Representation and Voice Desires

There is new evidence that employees, when asked what issues they want to express through enhanced voice mechanisms, express diverse interests (Kochan, Kelly, Yang, and Kimball 2018). What is not clear is whether the diversity of wants expressed by employees now is really more diverse than what employees would have said if they had been similarly surveyed in the 1920s, before unionism became more common in the United States. Maybe due to changes in the nature of work or the diverse demographic composition of the workforce, employee interests truly have become more diverse. But perhaps a lack of experience with voice mechanisms among current employees contributes to diverse wants, and if current collective representation forms were to become regularized and more familiar to employees, greater convergence in employee wants might emerge.

Similarities in Labor Relations Systems in the United States and Emerging Countries

As mentioned earlier, my recognition of the importance of collective presentation in the United States was in large part stimulated by my awareness of the central role that NGOs play in emerging countries. (For evidence on the latter, see Eaton, Schurman, and Chen 2017.) On the one hand, recognition of the growing influence of—and variety of—collective representation organizations and processes in the United States can be viewed as a positive development in that these new forms of representation provide a potential vehicle by which the "representation gap" that exists in the United States can be closed (Freeman and Rogers 2006). Yet there is a more worrisome aspect to this development—namely, that it is one among several ways by which the labor relations system in the United States is becoming more similar to the labor relations systems that exist in emerging countries.

Other similarities the United States has with emerging countries are the very low level of union membership in the private sector (and, consequently, collective bargaining where it exists at all is found in the public or non-for-profit sectors) and the fact that labor relations mostly involves political rather than private action. While I, and I suspect many of you, earlier had hoped that the American labor relations system would become more similar to the systems found in Europe through the expansion of labor–management partnerships and statutory due process protections at the workplace, instead the United States in its labor relations is becoming more similar to a third-world country.

The Future

I am reluctant to forecast whether the growing collective representation processes will reverse this trend and stimulate collective bargaining or become so institutionalized as to provide a widespread means of employee voice. When making predictions about the evolution of labor relations, I keep in mind that no academic in our field in the 1920s forecasted the explosive growth in unionism that took place in the 1930s. While our ability to forecast may be limited, I do think it is imperative that we recognize the serious nature of this change in the form through which employees are trying to influence the terms and conditions of work, and we should shift our research and teaching accordingly. The activities of LERA also should respond to this development.

References

Brudney, James J. 2016. "Decent Labour Standards in Corporate Supply Chains: The Immokalee Workers Model. In J. Owens and L. Howe, Eds., *Temporary Labour Migration in the Global Era.* London: Hart Publishing.

Eaton, Adrienne E., Susan J. Schurman, and Martha A. Chen, Eds. 2017. *Informal Workers and Collective Action: A Global Perspective.* Ithaca, NY: ILR Press.

Fine, Janice. 2006. *Worker Centers: Organizing Communities at the Edge of the Dream.* Ithaca, NY: ILR Press.

Freeman, Richard, and Casey Ichniowski, Eds. 1988. *When Public Sector Workers Unionize* (NBER Conference Volume). Chicago, IL: University of Chicago Press.

Freeman, Richard, and Joel Rogers. 2006. *What Workers Want.* Ithaca, NY: ILR Press.

Katz, Harry C., Thomas A. Kochan, and Alexander J.S. Colvin. 2015. *Labor Relations in a Globalizing World.* Ithaca, NY: ILR Press.

Katz, Harry C., Thomas A. Kochan, and Alexander J.S. Colvin. 2017. *An Introduction to U.S. Collective Bargaining and Labor Relations,* 5th ed. Ithaca, NY: ILR Press.

Kochan, Thomas A., Erin L. Kelly, Duanyi Yang, and William Kimball. 2018, "Worker Voice in America: A Current Assessment and Exploration of Options." MIT Working Paper. Presented at the LERA National Meeting, Baltimore, June.

Maffie, Michael. 2018a. *Logging Out: Conflict and Behavioral Responses in the Rideshare Industry.* Presented at the LERA National Meeting, Baltimore, June.

Maffie, Michael. 2018b. *Three Essays on Employment Relations in the Rideshare Industry.* Unpublished PhD thesis, ILR School, Cornell University, Ithaca, NY.

Marquis, Susan L. 2017, *I Am Not a Tractor: How Florida Farmworkers Took On the Fast Food Giants and Won.* Ithaca, NY: ILR Press.

Organizational Culture Under Cross-Border M&As: Authenticity, Integration, and Downstream Effect on Employee Turnover

YI WU
HENGKE FU
ROBIN MENGXI YANG
School of Economics and Management, Tsinghua University

MINGWEI LIU
School of Management and Labor Relations, Rutgers University

Abstract

This research tries to address the organizational culture authenticity problem in the context of cross-border mergers and acquisitions through in-depth interviews. A grounded theory is adopted and results in the following conclusions: (1) organizational culture authenticity indeed exist in organizational context but connotes meanings beyond the consistency and originality denotations, i.e., the extent to which an organizational culture being true to itself and its employees; (2) four dimensions of organizational culture authenticity have been identified: sustainability, credibility, spirituality, and identity; (3) three coping strategies used by employees with organizational shifts have been identified: rationalizing, practicing, and participating

Introduction

Giant global firms are facing problems in recruiting and retaining talented employees in emerging markets, particularly senior and experienced ones. Take China as an example. As the Chinese government stops offering political and economic benefits to foreign companies, while local firms emerge as strong competitors in the market, more and more job seekers prefer to work in local firms. A report published in 2014 pointed out that over 50% of Chinese senior executives who had been dismissed from foreign companies subsequently chose to work at local Chinese companies. This phenomenon becomes even more prominent when foreign companies acquire local firms. Many original employees decide to leave the company, turn to local firms, or even start their own business. They stated that they could not adapt to the new organizational culture and institutional procedure: "The [new] company is not the one I worked for. It is inauthentic," said a senior executive who left the acquired company and started his own business.

Organizational culture is the summarization of company-specific characterized values, group consciousness, behavioral norms, and thinking patterns that are generated during the long-term managerial practices and are accepted and followed by all the members (Schein, 1984). Cross-border mergers and acquisitions (M&As) amplify the significance of organizational culture by merging two previously distinct organizational cultures during a very short time. This situation can cause confusion among employees, as they could not work in the previous way (i.e., "inauthentic"). The new and previous values, rules, and relationships often conflict with each other, causing workers who have connections to the former organizational culture to leave the new organization. Therefore, exploring why this happens is of great importance in providing guidelines for companies undergoing such a huge transformation.

Corresponding author's address: Mingwei Liu, School of Management and Labor Relations, Rutgers University, New Brunswick, NJ 08904

However, although organizational culture, cross-border M&A, and employee turnover are all primary research topics in management, little research could be referred to address the very specific problem here. The current research places too much emphasis on the static perspective of organizational culture and does not explore the detailed mechanisms of how culture and identity conflict and integrate after M&As from the employee perceptual level versus the strategic level. The study of how such circumstances drive organizational cultural identification and employee turnover provides leave rich opportunities for new research.

Inspired by the practical and theoretical values of this topic, we want to address the problem by proposing three main research questions:

First, why do employees tend to leave the company that is merged or acquired by a company with a distinctive organizational culture? Does the term "organizational culture authenticity" really exist, and what is the dimensional structure?

Second, could a general process model be constructed to illustrate the entire cultural transformation process at both the employee level and the organizational level? What is the role of the "organizational culture authenticity" in this process, and how does it evolve?

Finally, what are the coping strategies adopted by employees when facing a huge organizational culture shift?

Literature Review

Cultural Integration in Cross-Border M&As

If each company has its own organizational culture, then an important downstream consequence of a cross-border M&A is that both companies face clashes of the two existing organizational cultures. A more formal definition comes from Berry (1980), who specifically points out that cultural integration is characterized by contact of two independent cultural groups, a systematic procedure containing contact, conflict and adaptation phases, and manifestation at both the individual and collective levels. Table 1 summarizes some of the literature on the relationship between cultural integration and M&As.

Table 1. Research on the Relationship Between Cultural Integration and M&As

Authors	Findings
Andre and Laurent (1983)	Managers from international companies think cross-cultural cooperation bring negative results.
Hapeslagh and Jemison (1991)	Corporate cultural integration is the foundation of operational integration.
Coopers and Lybrand (1992)	A case study on more than 100 M&A cases found that 85% of the CEOs believed that the differences between organizational cultures and styles resulted in M&A failure.
Kotter and Heskett (1997)	Companies that have good cultural integration achieve a 68.2% increase in income, whereas those with bad integration see only 16.6%.
Tetenbaum et al. (1999)	Avoiding organizational culture shock is key to a successful M&A.

Match-Oriented View vs. Process-Oriented View of Cultural Integration

The match-oriented view emphasizes that the two entities should share both a strategic direction and an organizational culture. Strategic fit focuses on complementarity between M&A enterprises in aspects such as products, markets, and industries. Therefore, a better resolution might be paying attention to how a beneficial cultural integration happens, i.e., a process view. Jemison and Sitkin (1986) point out that integration measures after an M&A have a greater impact than ad hoc matching itself. Alok and Chakrabarti (1990) also think that a successful M&A depends on strategic and organizational integration, with cultural integration as an important focus (Malekzadeh and Nahavandi 1990; Cartwright and Cooper 1993).

Acculturation Modes

From the dynamic perspective of typology, by observing and researching both acquiring and acquired companies, Nahavandi and Malekzadeh (1988) propose a model on acculturation mode selection on the basis of industry relevance, respective organizational culture identification, and the extent of integration. Specifically, this model adapts that of Berry (1980) but emphasizes in a proactive way how acquirers could subjectively put which acculturation modes into practice. Four types similar to those of Berry are also presented: integration, separation, assimilation, and deculturation. Furthermore, Weber et al. (2011) take national culture into consideration and embed Hofstede's (1980) national cultural dimensional theory into their research on acculturation mode selection. They believe that using these acculturation strategies can achieve the best M&A performance (see Table 2 for more details).

Table 2. Cross-Border M&A Acculturation Model (Weber et al. 2011)

National Cultural Attributes of M&A Enterprises	Level	Integration Method	Symbiotic integration Strategy
Circumventing uncertainty	High	All absorption and part retention	Absorption
	Low	Part absorption and all retention	Retention
Masculine temperament vs. feminine temperament	High	All absorption and part retention	Absorption
	Low	Part absorption and all retention	Retention
Power distance	High	All absorption and part retention	Absorption
	Low	Part absorption and all retention	Retention
Individualism vs. collectivism	High	All absorption and part retention	Absorption
	Low	Part absorption and all retention	Retention

Employee Turnover

Generally, there are two kinds of definitions of employee turnover. On the one hand, employee turnover is defined, in a broad sense, as "an individual's condition change as a member of the organization" (Price 2001). Following this definition, the hiring, promotion, demotion, transfer, and leaving behaviors should all be included in the realm of this definition. On the other hand, a narrower definition proposed by Mobley (1982) argues that employee turnover refers to the process in which individuals who receive material benefits from organizations terminate their relations with the organization as one of its members.

Theory

Authenticity is an important part of contemporary life. Beverland and Farrelly (2010) generally defined authenticity as "the genuine, real and true things." In this context, we reviewed the extant literature and summarized the relevant definitions on subtypes of authenticity. In general, extant research has divergent definitions and measurements of brand authenticity (e.g., Bruhn et al. 2012; Napoli et al. 2014; Morhart et al. 2015) and has not achieved a final conclusion.

There are differences in the scales proposed by Napoli et al. (2014) and Morhart et al. (2015). Napoli and his colleagues believe that consumers have different authenticity experiences. According to their own interests and knowledge, consumers use a series of cues to evaluate the authenticity of objective things. Judgments about authenticity can be based on either indexical cues (the physical connection between things and time) or iconic cues (things or events are a reasonable reconstructions of the past). In contrast, Morhart et al. (2015) think that brand authenticity perception reflects the result of objective reality (objective authenticity), subjective psychological association (constructed authenticity), and brand-related existentialist motives (existential authenticity). Following this principle, they extracted four brand authenticity dimensions: continuity, credibility, integrity, and symbolism, all reflecting objective, constructed, and existential authenticity at the same time.

Methodology

The primary purpose of this research is to explore whether the construct of authenticity exists theoretically in the organizational context, as well as the process in which it drives employees' turnover tendencies. Therefore, this study adopts qualitative research methods, which are appropriate in eliciting deeper insights on all related aspects. In-depth interviews with employees and analysis of second-hand documents analyses were carried out. The text and data, originally in Chinese, were analyzed via content analysis and then translated into English.

Company Selection

We selected a business-to-business Chinese company that provides power and electronic devices and services as the case company (the company is referred to as EV). This company was set up by a major Chinese local company in the 1990s and around the year 2000 was acquired as a subsidiary of an international company based in the United States. In 2016, the company was again sold to another American firm, leading to numerous employees leaving the company.

In-Depth Interviews

In-depth interviews are one of the most primary, practical, and worthwhile means to explore people's thoughts, attitudes, knowledge, and feelings on a specific occurrence (Groom and Lettlemore 2012). The interview type adopted in this research was a semi-structured interview.

Specifically, an outline was constructed for the interview. It consisted of three parts: overall perception and attitude toward organizational culture change, conflicts emerging in daily life and ways to address them, and interviewees' turnover tendencies in relation to those factors (see appendix for details). Generally, all the interviews started with a clarification that the purpose was to conduct academic analyses, and confidentiality was highly ensured. The interviewers were conducted one-on-one, either face-to-face or via telephone. Fifteen workers were interviewed: seven were senior executives, while eight were lower-level or newer employees who had been with the company for no more than two years. Half of the interviewees had left the company at the time of the interview. Four of the interviewed employees were female. See Table 3 for a summary of interviewee demographics.

Table 3. Interviewee Demographics and Interview Type

ID	Code	Gender	Years with Company	Left the Company	Department	Senior Executive	Interview Type
1	N1	Male	8	No	Channels	No	Face to Face
2	N2	Male	2	No	Sales	No	Face to Face
3	N3	Male	2	No	Sales	No	Face to Face
4	N4	Male	14	Yes	Sales	No	Face to Face
5	N5	Female	2	No	Sales	No	Face to Face
6	N6	Male	14	No	Product	No	Face to Face
7	N7	Female	8	Yes	Sales	No	Face to Face
8	N8	Male	8	No	Sales	No	Phone
9	N9	Male	14	No	Sales	Yes	Face to Face
10	H1	Male	13	Yes	Sales	Yes	Phone
11	H2	Male	20	Yes	Sales	Yes	Phone
12	H3	Male	20	Yes	Marketing	Yes	Face to Face
13	H4	Female	20	No	Product operation	Yes	Phone
14	H5	Female	20	Yes	Finance	Yes	Phone
15	H6	Male	20	Yes	Telecom operator, marketing	Yes	Phone

Second-Hand Documents

As a complementary information source, news reports were reviewed on the two major M&A events of the company to obtain a thorough description of how the company's organizational culture evolved over the past 20 years. Memoirs and books on the two M&A events written by senior executive witnesses were also examined to understand the psychological transformation of the employees.

Data Analysis

The collation, analysis, and interpretation of qualitative data followed a three-step procedure proposed by Spiggle (1994):

1. Classification and coding. Interview transcripts were examined iteratively to obtain full knowledge of the details and grasp the whole framework (Holbrook and O'Shaughnessy 1988). Then the data were initially classified and coded by hand.
2. Data analysis. The analysis procedure consisted of categorization, abstraction, comparison, dimensionalization, integration, iteration, and refutation.
3. Interpretation. Following the traditions of hermeneutics (Spiggle 1994) and the principles of grounded theory (Egan 2002), a basic theory was induced about the authenticity of the organizational culture context and its downstream consequences on employee turnover. Also a triangulation analysis was conducted until theoretical saturation was achieved.

Results

The transcripts were scrutinized to answer the following questions:

1. What opinions do employees hold about changes and transition in organizational culture, especially during the period after being acquired? In other words, does authenticity really exist in the organizational behavior research area?
2. From a theoretical perspective, what is the basic definition of organizational culture authenticity? Can a certain dimensionality be identified and summarized on the basis of the qualitative data?

3. How is employee turnover behavior affected by organizational culture authenticity? Can we delineate the entire process mechanism in a more detailed way by adding employee-level characteristics?
4. What steps could companies take to smoothly manage the post-acquisition period?

Authenticity Connotation in Organizational Culture Transition

Change as a Company's Nature

Generally, employees of EV, whether still employed with the company or not, believed that a change in organizational culture is certain to occur in an M&A. For example,

> In my opinion, if a company decides to acquire another company, it must have its own concerns and expectations, rather than maintaining the status quo of the acquired company. The company conducts the acquisition not for fun, but because it thinks the acquired company will generate values in the future. There must be a reason. To fulfill goals like this, the acquiring company is sure to reform the acquired company. … It's impossible to remain unchanged, unless the acquiring company just wants to buy but intends to do nothing about the acquired company. But it is so unrealistic. *(N5 interview excerpt)*

Equivocality in an M&A

In terms of the specific culture changes after the recent acquisition, employees primarily made one key point: VT (the acquiring company) tried to change the main marketing performance appraisal indicator in EV (the acquired company) from profit margin to profit amount. But overall, our interviewees held an equivocal attitude about what had occurred, mainly because they thought more time was needed to examine the effectiveness of the changes, and because each organizational culture (VT and EV) had its pros and cons. As one interviewee said,

> Comparing the previous and current organizational cultures, I think it is hard to tell which one is better. This is what the senior executives should balance. … I know there must be some time for transition. … It's hard to verify in a very short time, as we don't know whether follow-up measures will take place and whether conflicts will emerge afterward. *(N5 interview excerpt)*

Professionalism as an Employee Principle

More important, the attitudes from both senior and staff-level employees emphasize professionalism. To be more specific, they said that they should actively adapt to the changing trend after an M&A and regard professionalism as the basic principle to follow:

> To a person who has entered a company, if wanting to stay and to continue working, he or she has to accept the changes, no matter how the company will change. *(N5 interview excerpt)*

Dimensionality of Organizational Culture Authenticity

Sustainability

The prevalent concern of participants about the organizational culture, especially after the M&A, was the sustainable development of both the company performance and their vocational life. In reality, nearly every interviewee mentions his or her willingness to earn higher pay, enhance work efficiency, and boost the company's competitive advantage by taking opportunity of the M&A events. And that's why not all of them held negative attitudes during the turbulent transition. For example, an employee who had been at the company for 14 years said that he currently has a conservative opinion about the future, as he regards the status quo more as a temporary situation and argues that anything (bad or good) could happen in the future:

> Changes I have been feeling? I think the most obvious one might be that we are in a relevant changing procedure. So everything is not so stable, as we need such a transforming period. Thus there are so many uncertainty factors, whether

> the whole organizational structure, our future emphasis, the development direction, etc. … I am concerned because of this uncertainty, but it is hard to evaluate whether it's good or not. … Also … I cannot judge [whether] the team adjustment will bring the company better development, make it [maintain] what it is, or lead itself to even bad consequences. *(N4 interview excerpt)*

Instead of just expressing concerns about the impact brought about by the M&A, employees noticed which specific changes in organizational culture took place after the acquisition and judged for themselves whether such changes would be positive in the long run. As long as the organizational culture consistently promotes the company, employees felt that both the acquisition and the new organizational culture were good.

Credibility

Sense of Belongingness. When an acquisition happens, employees also claimed that their relationship with the company is dissociated, i.e., their sense of belongingness is decreasing. These emotionally detached feelings are caused by multiple reasons. First, some people mention that they treasured the time working in the company because they grew up knowing the brand and the company. Even young employees who just joined the company asserted that the work gave them a sense of self-fulfillment, identification, and belonging. For example:

> *N4:* I think influences [brought about from the M&A] are not so significant, especially to the essence of our jobs [not being] affected. The most salient point lies in the loss of EV's brand during the M&A process, which made us lose faith at that time, more or less.
>
> *Interviewer:* What's your understanding of the EV brand?
>
> *N4:* At that time, EV reached the summit of the industry.
>
> *Interviewer:* So it was the top one in your industry, right?
>
> *N4:* At that time, yes. I was working with it for 14 years. I experienced a lot during this time and had no hesitation about what EV did achieve.
>
> *Interviewer:* Besides the quality standards of EV, what other emotional components does EV possess?
>
> *N4:* You mean my own opinion? Briefly, I [grew up] with this brand and learned many things. After leaving [the company], I started a business by myself. The capabilities were almost the same [as] when [I was at] EV. Although I expected … unfavorable effects of acquisition and started to consider other brands' products, … I feel attached to this brand.… I spent so many years in this company, 14 years. And it's my first job.

Sense of Trustworthiness. Another aspect of credibility can be represented by a sense of trustworthiness, which can be described as the extent of employees' faith in their colleagues and the company—in other words, whether employees are willing to act in accordance with their colleagues' and the company's requirements, behaviors, and decisions. For example:

> [After our company was acquired by EV], initially I did not see too huge [of a] change. It was because my department did not undergo a big change. After [the] acquisition, EV only assigned three people to manage our company: one was the CEO, an ABC who was quite American; a pure American who was the CFO, and a financial director. Only these three. …The other employees remained the same. So it was sending us a signal that the new company agreed

> [with] what we had done before, and that it trusted ... that we could create values. I think this was the most important thing at that time. *(N5 interview excerpt)*

Spirituality

Implemented Value Proposition. A company is characterized by certain organizational culture values at each stage of its development. Specifically, company WH focused on an aggressive culture orientation; employees of the company went all out for their goals and tried their best to satisfy customers' needs and demands. Work was emphasized as more important than individuals and family. However, during the EV period, family and health were claimed to outweigh work obligations, causing employees to be less motivated to work and the company to lose certain market opportunities. After the acquisition, VT re-emphasized a focus on business and work. But this was different from the culture values initiated by WH, where organizational values were presented to employees so strongly that they were regarded more as requirements than edification. However, it can be inferred that value orientation plays a key role in organizational culture because some values still exert influence even after decades of have passed. For example:

> *Interviewer:* Do you think in late 2010, [WH's] culture still exists in EV?
>
> *H2:* Yes, I think so. We were always talking about WH's culture, so often that sometimes the senior executives of EV even felt uncomfortable. They said that you [have] been [with] EV for such a long time, why [are you] always talking about WH? In reality, it was WH's culture that affected the group of us so deeply, ... but what were the changes? Two major ones. First, the plan management. What we did [was] all planned and administrated. And then the financial management. Entirely following the Fortune 500 way. Further, everything we did had to follow the rules. Foreign companies have their own business rules, ethics, standards, etc. But down deep in our heart, our working styles were WH's, our values were WH's.

However, although the impact of organizational culture values is significant, the realization of its consequences requires down-to-the-end implementation, but not just talking about the values. A number of interviewees, including the younger ones who had been with the company for two years or less, said they could not ascertain the company's values. Such unfamiliarity leads (young) employees to be indifferent to organizational culture values or even think they are useless. Accordingly, organizational culture values should be reflected in the details of daily operation.

> *Interviewer:* When did you find such changes (in cultural orientation)?
>
> *H3:* Some key events. For instance, in WH it was common to see executives beat employees. But after being acquired by EV, people who [engaged in] such behaviors were fired or demoted. Then [the] executives stopped doing violence to subordinates.

Moral Integrity. The other key—but less mentioned component of spirituality—is that employees care about whether the business operations "overstep the boundary"—in other words, whether their colleagues exhibit behaviors that are not a virtue. When interviewees talked about such issues, their negative attitudes were expressed. For example:

> There are some obvious changes. ... Now all the operational activities [are] marketed-oriented. So there is not either black or white. As long as laws and regulations allow, there are some flexible shifting approaches. *(N9 interview extract)*
>
> With EV's brand status improving, there were some spaces for rent-seeking. Also for rather a long time, power was not under regulation. So the brand

> started to decay from inside. The internal issues were getting worse and worse. When EV decided to sell us out, some problems were quite prominent. To me, the rooted problem happened after [the] distribution network had been constructed. Before that, local offices were not allowed to [be involved with] distribution issues. That was to [maintain] separation of power. But afterwards, … local power was not curbed and regulated by the company, causing power abuse. Many employees used such power to seek private benefits. Slight negligence may lead to great disaster. … The company was destroyed from inside. *(N9 interview extract)*

Identity

Self-Embodiment. Employees try to express their existing or desired identities when they work for the company or the brand. Thus, the concept of self-embodiment is presented here to denote whether employees' intentions to develop better selves could be reflected or even facilitated by the company's organizational culture. Generally, such embodiment can be summarized as two types from the interviewees' responses. On the one hand, employees proactively seek ways to push themselves to achieve subjective initiative to a greater extent, i.e., self-enhancement. To be more specific, when asked about how they feel after VT introduced an intense reporting system to regulate daily operation, most employees didn't perceive it as extra requirement or workload, but rather an external motivation to boost efforts and discourage laziness. They seem to be more involved in an organizational culture that values effort and work. For example:

> *Interviewer:* Do you think the current organizational culture, I mean, the work-centric mode, … lowers your enthusiasm for work?
>
> *N8:* I don't think so. … The work-centric mode [puts a] heavier burden [on] me. But people need external pressure, to some extent. It is caused by human nature, because everyone might be lazy to some extent. Actually, people who deal with marketing issues should not be lazy, because delay leads to losing of opportunities. People who are self-disciplined might not be affected by what the company emphasizes. But to be honest, laziness generates in the non-work-centric environment, which makes employees start to dawdle away their time.

Company/Brand Symbolism. Unlike self-embodiment, which describes the proactive willingness of employees to explore similar connotations in organizational culture in support of forming their own identities (both existing and desired ones), company/brand symbolism refers to the extent of how elements that characterize the company or the brand could be readily used by employees to symbolize themselves. This was seen in the interviewees' unfavorable attitudes toward the newly changed brand identity components, in terms of both intangible (e.g., faith or belief in company/brand) and tangible (e.g., company/brand logo).

Process of Organizational Culture Authenticity Affecting Employee Turnover

Generation of Organizational Culture Authenticity Perception

A Bidirectional Mechanism of Organizational Culture Permeation. The international M&A context adopted in this research provides an ideal situation to study how organizational culture transmits across the company in that it presents how old and new organizational cultures clash drastically in a short period, thus enabling a comparison. Unable to smoothly adapt to the new organizational culture probably leads to inauthentic perception in employees. After examining respondents' descriptions of how the new organizational culture substitutes for the old one in the two acquisitions, both top-down and bottom-up routes were identified.

On the one hand, abstract cultural, moral, and even general business concepts or beliefs should be transmitted in a top-down way. In other words, symbolic or spiritual organizational culture concepts should be felt or perceived by employees rather than being told to them. For example, a senior executive (*H5*) who spent 20 years in the company mentioned a specific situation that impressed on her how the organizational culture was different after WH was sold to EV (as illustrated in her comments later). Employees probably cannot achieve deep and concrete feelings and thoughts about what exactly these new organizational culture orientations are. For instance, employees who had been in the company for no longer than 10 years consistently stated that organizational culture is a hollow thing referred to by senior executives in conferences but never seen in their work (according to comments by N1, N3, N5, and others).

On the other hand, respondents specifically noted that when the acquiring company tried to reframe or adjust the business process, it was better to make decisions based on the actual situation, rather than just follow outsiders' (e.g., external consultants) suggestions or advice—i.e., the top-down approach. Some respondents even mentioned the word "empowerment" several times to express their desire to remain doing what they usually do in daily operation or their willingness to let their own opinions on the company's business be heard or accepted, as employees believe they are the ones who are familiar with and have expertise in the acquired business. Here, self-values were repeatedly referred to, reflecting employees' wish to be respected professionally.

The Connecting Role of the Executive Level. In the bidirectional mechanism of organizational culture permeation, the key role of the executive level should not be ignored. First, from a functional perspective, executives are the ones who both receive cultural values and business orientations from the parent company and transmit such concepts to their subordinates. A junior staff member (*N5*) held the opinion that the direct superior leaders have a more critical impact on them than the overall company atmosphere itself. *H4* said that the company should be certain the executive level understands and accepts the company's concepts of organizational culture and business operation, and then further take the lead as examples to their subordinates. *H5* added that, without executives, the company could not perform effectively, even if the organizational culture and business orientation are ideal.

Second, driven by the emotional motivation to protect their own teams, executives themselves had the innate tendency to shoulder the responsibility as a bridge to carry out the organizational culture and to prevent their team members from being negatively affected by the turbulent organizational atmosphere in the acquisition. A detailed example of an executive's negative feeling when she received no feedback, and even rejection from the company, is found in this statement:

> *H5:* People like us surely hope that we offer values to the company and grow with the company. Make a difference, I mean. Because you are in the company, [the] company becomes different. I think it is important. It is useless to *me* to stay in this company, as if everyone is obedient.
>
> *Interviewer:* When did you have such feelings and start to practice that?
>
> *H5:* I was not alone in the company. I had my own team. I was always taking the leading role [with] such attitudes. I also hoped they would not be affected by the side effect of acquisition. But after I tried everything, I could not protect them. Won't I feel sad?!

Employee Turnover

Factors Driving Employee Turnover. All the respondents were asked why they left the company or what would be the driving factors if they left the company. Such factors all emerged after the acquisition, illustrating the downstream impact of an organizational culture shift at the employee level. The responses were summarized to illustrate reasons that drove employees away from the company after acquisition:

> *Interviewer:* Do you think this overall change affects you [very] little?
>
> *N9:* Not really. You should say, it changes but I can accept such changes. I want to see whether such changes are effective.
>
> *Interviewer:* You want [to] try?

> *N9:* Yes, exactly. We cannot deal with things emotionally. We should respect the company's suggestions and guidance. If their advice really works well, it is our emotional rejection that stops us from changing.

Coping Strategies with Organizational Culture Shift. Three strategies that employees normally adopt to cope with organizational culture shift are summarized. Such strategies differ from authenticating strategies that help people process information to achieve authenticity. Organizational culture authenticity can be achieved if the organizational culture achieves sufficient acceptance of its four dimensions, while coping strategies could only attenuate the psychological discomfort brought about by sudden organizational culture shift and it side effect of turnover threat, but it could not affect or rationalize the information processing procedure to authenticate.

The first strategy is *rationalizing*. Consistent with what was described previously as "professionalism as employee principle," employees like to rationalize that organizational culture changes are normal. Consequently, they convince themselves that even unfavorable results of such shifts, such as leaving the company, are acceptable. For example:

> *H4:* I can leave the company at any time. I am getting old; 40 years old means that you are an old employee in the company. So I prepare myself to leave at any time. I believe that technically I don't have any problem. But from the perspective of cost, I've been working with the company for 20 years, so the company must pay me much more than other people. But does the company have to pay so much? Thus, psychologically I prepare well.
>
> *Interviewer:* So, subjectively, you don't intend to leave, but you have [formed] the psychological defense?
>
> *H4:* Yes, I can leave at any time.

The second strategy, *practicing*, refers to conformity to and implementation of the updated organizational culture beliefs and business operation rules enforced by the acquiring company. Unlike rationalization, which is related to psychological conviction, practicing is related more to passive behavioral performance. Respondents often said that they only wanted to fulfill what the company asked them to do, and desired no more in regard to a team atmosphere, sense of trustworthiness or belonging, or promotion potential. For example:

Finally, *participating* involves subjectively changing working styles and even learning extra skills or knowledge to better meet the new requirements enforced by the acquiring company. As emphasized by the respondents, employees proactively adjust to the new organizational atmosphere not only because they have the innate desire for professional development and opportunities, but also because they try to demonstrate their competence in a turbulent environment.

> *Interviewer:* Have you experienced any struggles, and what did you do to solve such problems?
>
> *H4:* Actually from WH to EV, I did not do too much. The acquiring company originated in the U.S., and I found that I had not spoken English for a long time. What I did was [go] to [an] English class in the evening, and also [take] courses set by our company, mainly to practice oral English. Furthermore, the company emphasized data and indices, so I read a lot of books on data processing to make sure that I worked efficiently when dealing with data issues. In addition, presentation style was also important in EV, so I tried to fulfill EV's requirements when I [gave] a presentation, to let others understand. Overall, you have to [make these adjustments] within one or two years. Any changes require adaptation.

Conclusion and Discussion

Conclusion

Our findings contribute to the understanding of employees' perspectives on organizational culture changes on the basis of authenticity theory. We find that in an organizational behavior context, employees do not insist on adhering to the previous organizational culture. Specifically, four dimensions are identified that consist of fundamental components of an authentic organizational culture: sustainability, credibility, spirituality, and identity (see Figure 1 for more details).

Figure 1. Dimensionality of Organizational Culture Authenticity

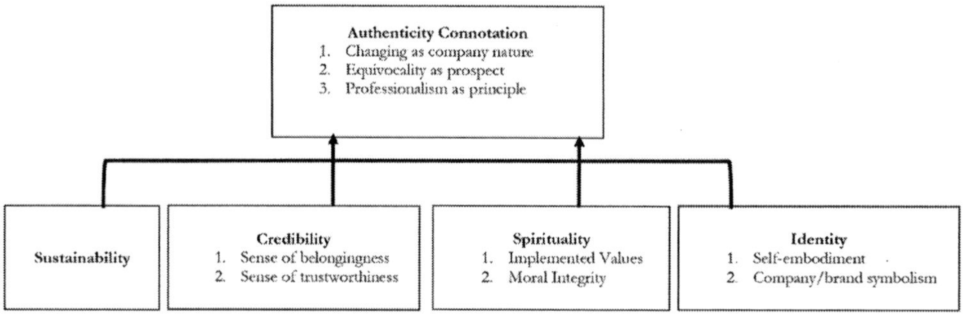

Furthermore, we try to interpret how organizational culture authenticity affects employee turnover by forming a process model from the theoretical basis of a cognitive and affective system of personality (CASP) theory, meaning that abstract and spiritual culture values and business operation reframing are evaluated in separate ways and should be adjusted in different ways. We find that, when the organizational culture changes, it follows a dual process to transmit the changes throughout the company—a process in which the executives serve as the link to communicate information and concepts.

Specifically, abstract organizational culture concepts should be transmitted from the top down, while business operation reconstruction should be from the bottom up. We also find that factors leading to employee turnover could be framed into two similar ways: (1) the more cognitive and tangible welfare and promotion process and (2) the more affective and intangible organizational culture-related process.

Three strategies adopted by employees in coping with such organizational culture shift are also identified: rationalizing, practicing and participating.

In articulating our findings, we provide here an overarching process framework for understanding employee perceptions of organizational culture authenticity. The findings demonstrate a sequential process of how organizational culture change affects authenticity perception and further results in employee turnover (see Figure 2 for more details).

Figure 2. Manifestation of Authenticity in Organizational Culture Shift

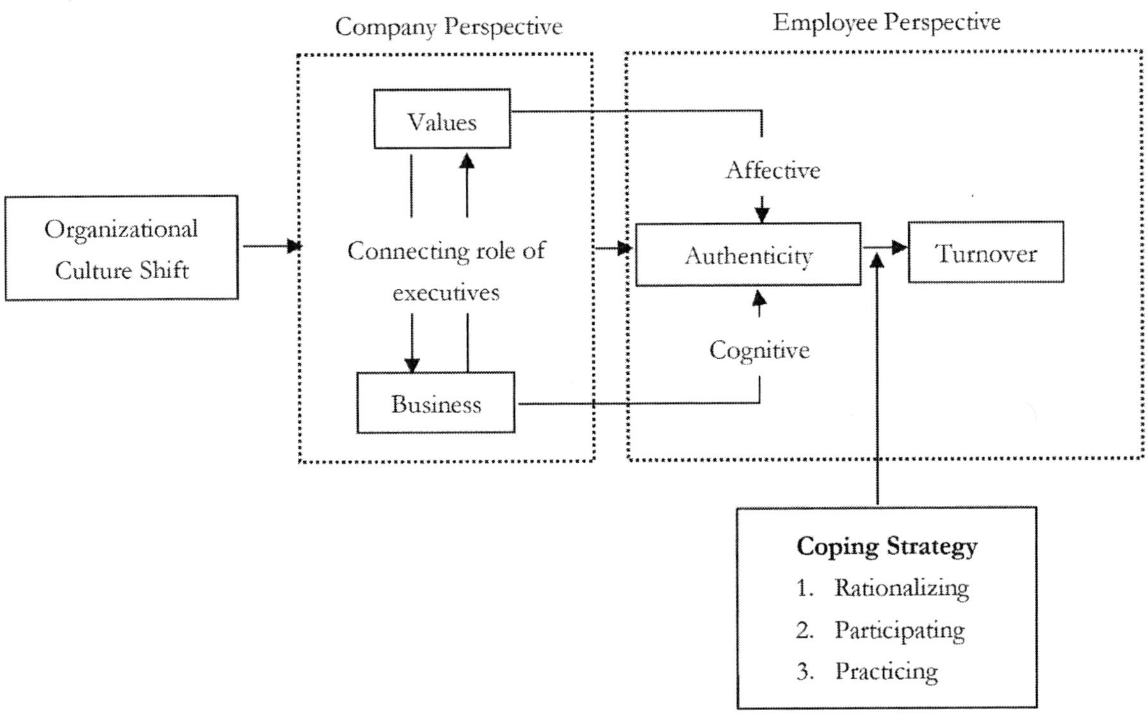

Managerial Implication and Limitations

First, for organizations experiencing substantial reform or turbulence, we offer a framework to evaluate and monitor the organizational culture change. Second, the bidirectional mechanism on how renewed organizational culture flows across the acquired company offers insights and guidelines for companies in a similar situation to more effectively manage the organizational culture transition.

This research is not without limitations. The major one lies in relying primarily on in-depth interviews as the research methodology. Accordingly, whether the definition, dimensional structure, and process model are reliable requires empirical examination. Furthermore, although the case selected for study probably reflects the problem of organizational culture shift and transition, future research should focus on extending the external validity of the conclusions drawn here. Finally, it will be useful to develop scales for organizational culture authenticity and examine its nomological validity.

References

Aaker, D. A., Stayman, D. M., & Hagerty, M. R. (1986). Warmth in advertising: Measurement, impact, and sequence effects. *Journal of Consumer Research, 12*(4), 365-381.

Ahammad, M. F., & Glaister, K. W. (2011). The double-edged effect of cultural distance on cross-border acquisition performance. *European Journal of International Management, 5*(4), 327-345.

Altman, J., & Bayer, S. A. (1990). Migration and distribution of two populations of hippocampal granule cell precursors during the perinatal and postnatal periods. *Journal of Comparative Neurology, 301*(3), 365-381.

Ashforth, B. E., Harrison, S. H., & Corley, K. G. (2008). Identification in organizations: An examination of four fundamental questions. *Journal of Management, 34*(3), 325-374.

Batra, R., & Ray, M. L. (1986). Affective responses mediating acceptance of advertising. *Journal of Consumer Research*, *13*(2), 234-249.

Burke, M. C., & Edell, J. A. (1989). The impact of feelings on ad-based affect and cognition. *Journal of Marketing Research*, 69-83.

Berry, J. W. (1980). Acculturation as varieties of adaptation. In A. M. Padilla (Ed.), Acculturation: Theory, models and some new findings (pp. 9–25). Boulder, CO: Westview Press.

Bartels, J., Douwes, R., Jong, M., & Pruyn, A. (2006). Organizational identification during a merger: Determinants of employees' expected identification with the new organization. *British Journal of Management*, *17*(S1).

Boxx, W. R., Odom, R. Y., & Dunn, M. G. (1991). Organizational values and value congruency and their impact on satisfaction, commitment, and cohesion: An empirical examination within the public sector. *Public Personnel Management*, *20*(2), 195-205.

Bass, B. M. (1995). Theory of transformational leadership redux. *The Leadership Quarterly*, *6*(4), 463-478.

Brouthers, K. D. (2002). Institutional, cultural and transaction cost influences on entry mode choice and performance. *Journal of international business studies*, *33*(2), 203-221.

Bluedorn, A. C. (1982). A unified model of turnover from organizations. *Human relations*, *35*(2), 135-153.

Beverland, M. B., & Farrelly, F. J. (2009). The quest for authenticity in consumption: Consumers' purposive choice of authentic cues to shape experienced outcomes. *Journal of Consumer Research*, *36*(5), 838-856.

Beverland, M. B., Lindgreen, A., & Vink, M. W. (2008). Projecting authenticity through advertising: Consumer judgments of advertisers' claims. *Journal of Advertising*, *37*(1), 5-15.

Bruhn, M., Schenmuller, V., and Schafer, D., et al. (2012). Brand Authenticity: Towards a deeper understanding of its conceptualization and measurement. *Advances in Consumer Research*, 40, 567-576.

Chatman, J. A. (1989). Improving interactional organizational research: A model of person-organization fit. *Academy of Management Review*, *14*(3), 333-349.

Chatman, J. A. (1989, August). Matching people and organizations: Selection and socialization in public accounting firms. In *Academy of Management Proceedings* (Vol. 1989, No. 1, pp. 199-203). Academy of Management.

Cartwright, S., & Cooper, C. L. (1993). The psychological impact of merger and acquisition on the individual: A study of building society managers. *Human Relations*, *46*(3), 327-347.

Cartwright, S., & Cooper, C. L. (1993). The role of culture compatibility in successful organizational marriage. *The Academy of Management Executive*, *7*(2), 57-70.

Cartwright, S., & Schoenberg, R. (2006). Thirty years of mergers and acquisitions research: Recent advances and future opportunities. *British Journal of Management*, *17*(S1).

Chatterjee, S., Lubatkin, M. H., Schweiger, D. M., & Weber, Y. (1992). Cultural differences and shareholder value in related mergers: Linking equity and human capital. *Strategic Management Journal*, *13*(5), 319-334.

Christensen, C. M., Alton, R., Rising, C., & Waldeck, A. (2011). The big idea: The new M&A playbook. *Harvard Business Review*, *89*(3), 48-57.

Cameron, K. S., and Quinn, R. E.. Diagnosing and changing organizational culture: Based on the competing values framework. *John Wiley & Sons*, 2005.

Epstein, S. (1994). Integration of the Cognitive and the Psychodynamic Unconscious. *American Psychologist*, 49(8), 709-724.

Eggers, F., O'Dwyer, M., Kraus, S., Vallaster, C., & Güldenberg, S. (2013). The impact of brand authenticity on brand trust and SME growth: A CEO perspective. *Journal of World Business*, *48*(3), 340-348.

Elsass, P. M., & Veiga, J. F. (1994). Acculturation in acquired organizations: A force-field perspective. *Human Relations*, *47*(4), 431-453.

Fairfield-Sonn, J. W., Ogilvie, J. R., & Del Vecchio, G. A. (2002). Mergers, acquisitions and long-term employee attitudes. *The Journal of Business and Economic Studies*, *8*(2), 1.

Guerrero, S. (2008). Changes in employees' attitudes at work following an acquisition: a comparative study by acquisition type. *Human Resource Management Journal*, *18*(3), 216-236.

Grayson, K., & Martinec, R. (2004). Consumer perceptions of iconicity and indexicality and their influence on assessments of authentic market offerings. *Journal of Consumer Research*, *31*(2), 296-312

Holbrook, M. B., & O'Shaughnessy, J. (1988). On the scientific status of consumer research and the need for an interpretive approach to studying consumption behavior. *Journal of Consumer Research*, *17*(4), 440-453.

Haspeslagh, P. C., & Jemison, D. B. (1991). Managing acquisitions: Creating value through corporate renewal (Vol. 416). *New York: Free Press.*

Harris, P. R., & Moran, R. T. (1991). Managing Cultural Differences: High-Performance Strategies for a New World of Business. *Houston.*

Jemison, D. B., & Sitkin, S. B. (1986). Corporate acquisitions: A process perspective. *Academy of Management Review*, *11*(1), 145-163.

King, D. R., Dalton, D. R., Daily, C. M., & Covin, J. G. (2004). Meta - analyses of post - acquisition performance: Indications of unidentified moderators. *Strategic Management Journal*, *25*(2), 187-200.

Kim, J., Baek, Y., & Choi, Y. H. (2012). The structural effects of metaphor-elicited cognitive and affective elaboration levels on attitude toward the ad. *Journal of Advertising*, *41*(2), 77-96.

Groom, N., & Lettlemore, J. (2012). Doing applied linguistics: A guide for students. *London: Routledge.*

Price, J. L. (2004). The development of a causal model of voluntary turnover. *Innovative Theory and Empirical Research on Employee Turnover*, 3-34.

Morhart, F., Malär, L., Guevremont, A., Girardin, F., & Grohmann, B. (2015). Brand authenticity: An integrative framework and measurement scale. *Journal of Consumer Psychology*, *25*(2), 200-218.

Napoli, J., Dickinson, S. J., Beverland, M. B., & Farrelly, F. (2014). Measuring consumer-based brand authenticity. *Journal of Business Research*, *67*(6), 1090-1098.

Newman, G. E., & Dhar, R. (2014). Authenticity is contagious: Brand essence and the original source of production. *Journal of Marketing Research*, *51*(3), 371-386.

Organ, D. W., & Konovsky, M. (1989). Cognitive versus affective determinants of organizational citizenship behavior. *Journal of Applied Psychology*, *74*(1), 157.

Resick, C. J., Baltes, B. B., & Shantz, C. W. (2007). Person-organization fit and work-related attitudes and decisions: examining interactive effects with job fit and conscientiousness. *Journal of Applied Psychology*, *92*(5), 1446.

Spiggle, S., Nguyen, H. T., & Caravella, M. (2012). More than fit: brand extension authenticity. *Journal of Marketing Research*, *49*(6), 967-983.

Weber, Y., Tarba, S. Y., & Reichel, A. (2011). A model of the influence of culture on integration approaches and international mergers and acquisitions performance. *International Studies of Management & Organization*, *41*(3), 9-24.

Appendix

Question List for Interview

Part 1. Overall Perception (20 minutes)

1. Please briefly describe the M&A event of your (previous) company (e.g., by what company, the origin of the merging company, etc.).
2. What changes did you feel after your company had been acquired by another company (e.g., overall atmosphere, organizational culture, etc.)? Please illustrate by givng examples of your daily work. What were the good changes? And what were the bad ones?

3. In your opinion, what were the values and norms of the company before M&A? Did you identify with them? What were they after the M&A? Did you still identify?
4. In sum, do you think the company after the M&A was still the company you worked for? In other words, is it authentic? Could you give me several examples of why it is still authentic?
5. At that time, in order to keep the company authentic after the M&A, what aspects do you think could have been changed? What aspects should not have changed?

Part 2. Conflict Integration (25 minutes)

1. Because the company that acquired your (previous) company came from outside of China, the company had quite adifferent organizational culture or atmosphere from your (previous) company. Did you feel conflicts brought about from these differences? In what specific aspects?
2. Did the above conflict(s) alleviate afterward? How or how not)? What efforts did you contribute to this process? Did your efforts pay off?
3. In the process of addressing conflicts, do you still think the M&A brought good (or bad) changes to the overall atmosphere and organizational culture of the company? Please give me one or two examples.
4. When did you start to think the values and norms of the company after the M&A became acceptable? Were there any key events that inspired you to think so? Or, after going through such a procedure, why did you still refuse to accept them?
5. If you thought the cultural atmosphere of the company after the M&A was inauthentic, do you feel it became authentic after you experienced the conflict-addressing process? In what aspects and why? If you thought the cultural atmosphere was authentic, did you change your opinion after such a process? In what aspects and why?

Part 3. Turnover Tendency (15 minutes)

1. Did you want to leave the company after the M&A? Why? Was it related to the above-mentioned conflict?
2. If so, when did you make efforts to leave?
3. Did your opinion of the values and norms affect your decision? Why or why not?
4. Imagine that you were in the period when the company had just been acquired, based on your current knowledge and experience, would you make the same decision? Why or why not?

MBA-HR Versus MS-HR Graduates: Career Pathways Entered and the Knowledge and Skills Needed in Those Career Pathways

DAN KOYS
DePaul University

Abstract

I take job titles that MBA-HR graduates and MS-HR graduates reported just after graduation and categorize them into the career pathways used in the Occupational Information Network (O*NET). I then use O*NET's knowledge, skill, and workstyle importance ratings to measure components of the SHRM Competencies Model. ANOVA procedures show that the following components are significantly more important in HR than non-HR occupations: knowledge of HR, knowledge of administration, integrity, initiative, concern for others, speaking, active listening, social orientation, social perceptiveness, management of personnel, and negotiation. I use these results to suggest a market-focused HR curriculum.

Author's address: Department of Management & Entrepreneurship, DePaul University, 1 East Jackson Boulevard, DePaul Center 7024, Chicago, IL 60604

Explaining the Gender Salary Gap for the Educated

DAVID MCCLOUGH
Ohio Northern University

MARY ELLEN BENEDICT
Bowling Green State University

Abstract

Using National Survey of College Graduates (NSCG) data for different decades, the authors examine why the gender salary gap for education is persistent, despite increased educational attainment by women. The analysis accounts for occupation and explores occupational segregation decomposition. The findings indicate that occupation is an important determinant of the gap and that while overall occupational segregation has declined, within occupation pay gaps remain, particularly for women who remain in traditionally female jobs. Further, including a control for occupational feminization explains part of the gap as well. The analysis suggests that the integration of economic and social theory in future research will provide better assistance in determining policy to reduce the salary gap between educated women and men.

Introduction

Since the 1970s, expanded education and the workforce opportunities for women have contributed to narrowing the gender pay gap. As of 2013, women now earn more undergraduate degrees, Master degrees, and doctoral degrees than men (U. S. Department of Education 2012). Highly educated women perform jobs that seemed unlikely in the 1960s, yet still face a gender pay gap: 33% for those with a baccalaureate degree and 36% for those with an advanced degree (Schneider and Gold 2016).

Research examining the persistence of the pay gap identifies education and the movement of women into traditional male occupations as key factors contributing to the decline of the pay gap through the early 1990s (Blau and Kahn 2007). Since the early 1990s, the narrowing of the gap has slowed (see Figure 1). Research offers three explanations to explain the slow down since the early 1990s. One approach considers whether women exchange income for flexibility in scheduling to accommodate family obligations. A second approach examines how occupational segregation contributes to the gender pay gap, if women are "crowded" into a subset of lower paid occupations. A third approach considers how women are devalued in the workplace, thus feminization of occupations yields lower average salaries (Levanon et al. 2009; Mandel 2013).[1]

Using a unique dataset, we explore the relationship between salary and occupation of educated workers. The dataset has advantages over the U.S. Census data in that it focuses on college educated workers over several decades. The data can be used to examine how occupation affects the salary gap for the educated and whether occupational effects have changed over time. Given the ongoing persistence of the gender wage differential, we examine how changes in the proportion of women (or men) in an occupation explains the narrowing of the gender wage differential over time. In particular, we will examine how occupational segregation and feminization affects the gender wage gap for college educated workers.

Corresponding author's address: David McClough, James F. Dicke College of Business Administration, Ohio Northern University Ada, OH 45810

There are important policy implications of this research, especially during an era of promoting college education as a mechanism to secure higher salaries for men and women. The challenge facing society, policy-makers, and researchers is whether the goal is to close the pay gap, which is an objective motivated by relative gains in recent years, or the goal is to increase the earnings of women in an absolute sense. While educational attainment has contributed to absolute gains in the form of rising income, there is little support that educational gains by women have contributed to relative gains. This research acknowledges the ambiguity of the social good, yet seeks to understand the structure of the earnings differential. We recognize that absolute gains by women can result in relative gains to men, so the policy question that emerges may not be how to close the absolute gap, but rather, how to ensure that men and women have opportunities to pursue occupations that meet the abilities and preferences of the individual. A better understanding of the effects of occupational segregation and feminization of occupations will contribute to identifying inequities in opportunity that can inform meaningful policy

Educated Women, Salary, and Occupation

Earlier research has established that the decline of the gender wage differential began in the 1970s. George J. Borjas (1996) reported that wages of women were consistently around 0.6 of the wages of men, but then the gap began to close by more than 1% each year until 1990 when the wage ratio reached 0.72. Blau and Kahn (2007) confirm the constancy of the ratio up to the 1970s as well as sustained increase during the 1980s until the early 1990s. Benefitting from more data, Blau and Kahn note the slowing of the convergence during the second half of the 1990s and the increase again during the 2000s. Research controlling for endogeneity finds that occupation leads to lower salaries for women, despite educational attainment (Vella 1993). Blau and Kahn (2007) indicate that after controlling for occupation and industry, 41% of the pay gap remains unexplained. Relying on 2013 data from the Bureau of Labor Statistics (BLS), Emily Baxter (2015) states that it is "well known" that women working full-time, year round receive $0.78 for each dollar men receive. So when Claudia Goldin reports during an interview with Stephen Dubner (2016) that the ratio has increased from 0.59 in the 1970s to 0.77 by 2015, it is clear that the gap has narrowed when comparing full-time, year round workers. Despite those gains, the gender gap in earnings has not changed substantially since 2003 (see Figure 1).

Figure 1
Female to Male Earnings Ratio, 1979–2015

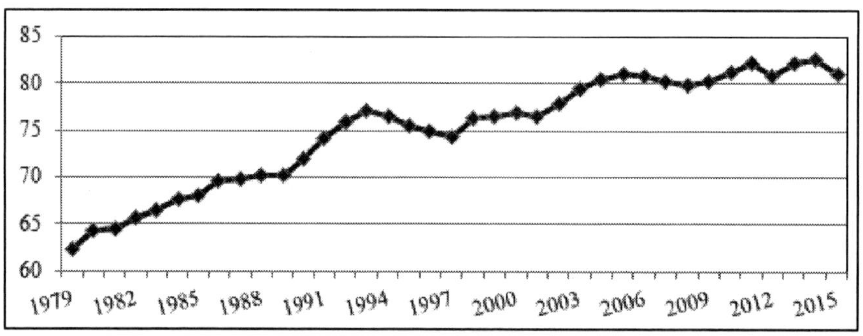

Data Source: Source: 1979-2015 Current Population Survey, U.S. Bureau of Labor Statistics.

While there seems to be no question that the gender wage differential has narrowed, there is less agreement regarding the forces driving the improvement. There seems to be a consensus that more women are working more hours for more years. BLS data reveal that the labor force participation rate for women maintained a steady increase since the 1940s before peaking just above 60% in March 2001. While the rise includes the decade of the 1980s, the civilian labor force participation rate for women has since receded to 56.8%. For comparison purposes, the rate for men has steadily declined from 86.4% in 1950 to 69.2% by 2016, so the narrowing of the gap likely results from more women working more years, but there is also the departure of less educated prime age men from the labor force, whose lower average income would contribute to narrowing the gender differential (Case and Deaton, 2017).

Andrew M. Gill and Duane E. Leigh (2000) used the National Longitudinal Study of the High School Class of 1972 (NLS72) and the National Longitudinal Study of Youth (NLSY) to examine how the relative growth in women's attendance at community colleges may explain the narrowing of the gender wage gap during the 1980s. The interest in community colleges is motivated by the emphasis on vocational training. NLSY data allows examination of changes in the gender wage gap over the 1985-90 period for 25-28 year olds and over 1989-94 for 29-32 year olds. The gender wage gap narrows for both periods. For the 25-28 cohort, decomposition reveals that work history explains 15% of the gap narrowing, while 38% of the narrowing of the gap is explained by an observed price effect. In other words, women are being paid for their skills and experience. Gill and Leigh find that attending two or four-year institutions and academic major explained 8.5% to 11% of the wage gap decline. Subsequent research has revealed that employment outcomes vary based on academic major (Benedict, McClough, and Hoag, 2012), so additional schooling can limit earning potential or enhance salary (McClough and Benedict, 2017).

The commitment to education by women continues to yield notable results. Women earn more undergraduate and graduate degrees than men and have done so for a decade (NCES, 2015). Overall, women earn 1.42 degrees for each male degree, yet the return on the investment in human capital is not clear. Claudia Goldin and Lawrence F. Katz (2002) argue that the emergence of women in the labor market and especially in professional graduate programs corresponds with legalization of oral contraceptives. They contend that the arrival of the contraceptive pill permitted women to dictate whether and when to have children, thus permitted women to invest in education, notably graduate professional education associated with higher earnings. Higher education, especially professional degrees, such as law and medical, affords women employment and lifestyle options not previously available as widely.

While women have posted noteworthy gains in educational attainment over the past four decades, the earnings ratio has not closed for educated women relative to similarly educated men. As shown in Figure 2, women still earn between 65 and 76 cents for every dollar men make, even for post-graduate degrees. Women earning a professional degree (medical or law degree) have the largest gender gap in salary, for each of the survey periods, while the gap is smallest among recipients of doctoral degrees.

Figure 2
Earnings Ratio for the Educated, 1993, 2003, and 2010

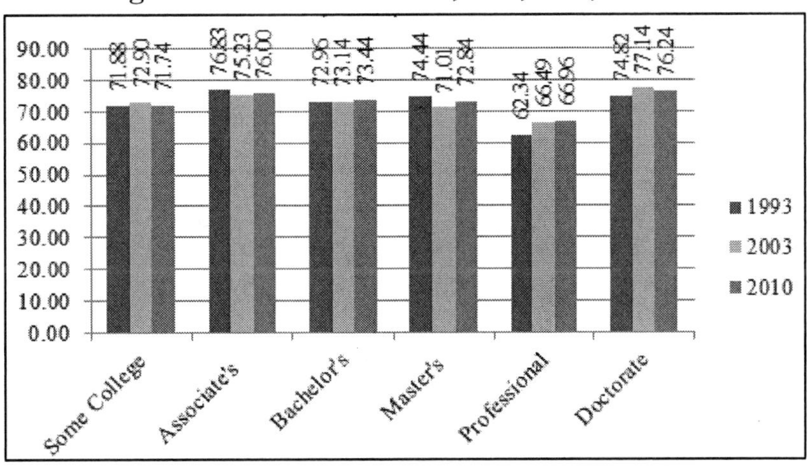

Data Source: U.S. Department of Commerce.

More education does not always result in higher salary. Linda Datcher Loury (1997) used NLS72 data and the High School and Beyond Senior Cohort (Class of 1980) to examine the narrowing of the gender wage differential during the period 1979-1986. Two data sets cover two separate periods, 1972 to 1979 to 1980 to 1986. Both samples are nearly all white, and the interviews occurred shortly after respondents completed college. The author contends that new graduates face current market conditions. The most notable difference across the two samples was the narrowing of the differential in choice of major. By 1986 more women were engineering majors and more men were education majors.

Loury (1997) finds that after controlling for academic major, market returns to women's skills (grades) increased relative to men. A smaller effect is also observed due to a change in the distribution of academic majors as men move into traditional majors of women and women enter academic majors historically male. This study concludes that as the distribution of majors changes, the greater effect on the pay gap is explained by market conditions that reward women's skills upon graduating college.

More recent research has found that the narrowing of the gap once limited to younger women has since extended to include more experienced women in the labor force as well (Blau and Kahn, 2007). Economic theory predicts that as the return on human capital rises that a rational actor will invest more in human capital. In this situation, however, the return accrues principally to work experience, which suggests that the investment will flow into work experience (Mincer and Polachek 1974), thus we may think of education as a necessary condition, but work experience is a sufficiency condition. We might then consider how decisions regarding marriage and children might be informed in relation to higher returns to work experience. Whereas the compensating differential argument asserts that women accept lower paying jobs to accommodate family obligations, we may want to consider the decision in reverse. Are highly educated women trading marriage and children for employment opportunities and income? Demographic evidence reveals that women are delaying child birth. Mathews and Hamilton (2009) use data from the Birth Data File, National Vital Statistics System, to find that the average age of first-time mothers increased from 21.4 years in 1970 to 25 years in 2006. It is convenient to observe that the increase appears to accommodate completion of college and any number of graduate or professional degrees that can be completed in one to three years.

After decades of examination, labor economists have made progress explaining the gender wage differential. However, narrowing of the gap implies a gap remains. Women have completely closed and, in fact, reversed the education gap and accumulated more experience, so why does a gap persist? While education is no longer the driving force behind narrowing the gap, the type of education matters. If more women pursue employment as elementary school teachers after college rather than as computer programmers, for example, years of schooling and degree attainment data will not explain the difference in salary, whereas occupation will.

Goldin and Katz (2016) use the case of pharmacists to illustrate that the degree of substitutability between men and women matters. In the case of pharmacists, two major changes have occurred in the pharmacy profession: independent pharmacies have by and large been replaced by large corporate alternatives and computer technology is employed extensively for management and communication. Pharmacy graduates enter the marketplace with identical training, experience, and skills. While there are distinct employment alternatives, the pharmacist occupation, whether in a hospital or retail, is essentially the same job. The one gender-based distinction once contributing to higher earnings for men was the greater tendency of men toward self-employment; however, the near extinction of independent pharmacies has greatly reduced the proportion of self-employed pharmacists. Goldin and Katz suggest that men and women are viewed as interchangeable variable inputs using computer technology that directs practice, so the wage differential for that occupation has disappeared.

For decades occupation has been employed to explain the gender differential (Blau and Hendricks 1979; England 1981 Cotter et al. 1995; Blau et al., 2012; Hegewisch and Hartmann, 2014; Alonso-Villar et al., 2012). Economic theory assumes that economic actors make choices that maximize utility, given their preferences and constraints. When an individual chooses to pursue a degree in science, for example, economists assume that the individual weighed the costs and benefits associated with science compared to alternatives. This approach inherently places the decision-making process of choosing a major or an occupation first, with resulting wages that follow. Woman may prefer certain occupations because of interests or external obligations. They also may be constrained by workplace or institutional discrimination. Empirically, economists have estimated preferences with choice models to account for the probability that one chooses to work (Heckman 1979) or one chooses a particular occupation (Vella 1993). To test for discrimination, studies decompose the salary regression; if the gender pay gap has a residual that cannot be explained, this portion of the gap is attributed to discrimination (Oaxaca 1979). More recently, research has acknowledged that the residual may not be discrimination entirely, but rather, there may be variables omitted from the wage regression. After all, testing models is limited by the data that exist.

Sociologists examine how societal values and stereotyping of gender roles affect the labor market. Queuing models (Levanon et al., 2009) array occupations by the rewards they offer. So, for example, if society values men more than women, men are selected to fill higher paying jobs ahead of women, who then occupy the remaining jobs that pay less (Peterson and Sapporta, 2004). In contrast, valuation models are built on an assumption that (gendered) cultural beliefs identify men as more productive than women. Accordingly, men are more valuable in the workplace. Men and women accept the gendered norm, which manifests itself with lower pay for occupations that are predominately female *because* women comprise those jobs (Levanon et al., 2009).[2]

To test these competing theories, sociologists investigate the level of occupational segregation in a society, which is represented by the level of occupational feminization. Karlin et al., (2002) tests for simultaneity between wages and occupational feminization, defined as the percentage of women in an occupation. Using Current Population Survey (CPS) from 1984-1991, the study indicates that wages are affected by feminization, but feminization is not a function of wages. Indeed, wages fall for both men and women in feminized occupations.

Studies provide strong evidence of devaluation (Levanon et al., 2009; Mandel, 2013), although some research finds that household obligations and job specialization are primary in explaining the gender pay gap. Javier Polaviega (2007) employs the European Social Survey with a focus on the Spanish sample. Using a Heckman selection model that estimates the individual's decision to work, he finds that the effect of occupational feminization disappears once job specialization is included in the wage equation. He concludes that occupational segregation plays only a minor role explaining the gender pay gap.

This study examines the gender salary differential for college educated workers. We examine three explanations for the persistence of the gender salary differential. The remainder of the study will proceed with description of the unique qualities of the dataset, followed by an empirical examination of each of the three possible explanations. The discussion of the results generates additional questions for future research.

The NSCG Data and Analysis

The National Survey of College Graduates (NSCG) sponsored by the National Science Foundation, surveys college graduates every decade.[3] The surveys produce large datasets comprised of 78,000 to nearly 150,000 respondents who received a degree in the United States. For most of the analysis, we include respondents who are identified as employed full-time and working in the United States for the 1993, 2003, and 2010 surveys. The data include demographic information, occupation categories, and job attitudes for 2003 and 2010 surveys. The 1993 dataset is the largest because 148,932 surveys were completed, compared to 100,402 and 100,000 for 2003 and 2010, respectively.[4]

To investigate the salary gap, we begin with an OLS regression model.

$$lnsal = X\beta + e \tag{1}$$

where *lnsal* is a vector of the natural logarithm of salary, X is a matrix of control variables, β is a vector of coefficients, and e is a vector of random error terms. The regressions employ NSCG sampling weights to provide estimates consistent with the relevant population. Due to nonconstant variance by gender and race groups, robust standard errors are used to correct for heteroscedasticity in the initial regression models.[5]

The salary regression models control for race, number of children, marital status, highest degree obtained, work experience, employer size, whether the employer is from the public or private sector, occupation, and region.[6] These are typical controls for salary regression models. The 1993 sample does not contain information on current job tenure or employer size.[7]

The NSCG includes variables that range from seven to 129 occupations. Many of the detailed occupations have zero or very low counts; therefore, we aggregate occupations. In earlier research (McClough and Benedict, 2017), fifteen occupational groupings seemed to work well and facilitate comparisons across the 1993, 2003, and 2010 surveys.

Table 1 presents results of the salary regressions. Due to the large number of observations, most coefficients are statistically significant at conventional levels of statistical significance. The F-statistics and Wald statistics indicate that the models fit very well. The adjusted R^2 values indicate more explanatory power for the 2010 data. This study focuses on the impact of occupation on the salary gap, so we do not list coefficient estimates for all variables.[8]

Table 1
Salary Regressions for the Full-Time Employed[a]

	1993		2003		2010	
	Unadjusted Gap=29.82%		**Unadjusted Gap=31.5%**		**Unadjusted Gap=25.9%**	
	(1)	**(2)**	**(3)**	**(4)**	**(5)**	**(6)**
Female	-0.232*** (0.003)	-0.127*** (0.003)	-0.271*** (0.006)	-0.176*** (0.006)	-0.222*** (0.009)	-0.135*** (0.009)
Occupations included	No	Yes	No	Yes	No	Yes
Adj. R^2	0.232	0.327	0.185	0.261	0.246	0.334
F-Statistic	1546.42***	1649.51***	414.51***	446.25***	259.74***	266.96**
N	106,772	106,772	52,667	52,667	47,664	47,664

[a]Regressions use NSCG data for the three years noted. Regressions include controls for demographic, family, employer, and 15 occupations. Contact authors for detailed regressions.

***statistical significance at 1 percent.

**statistical significance at 5 percent.

*statistical significance at 10 percent.

The unadjusted salary gap for full-time educated workers increased between 1993 and 2003 by 1.7 percentage points and declined between 2003 and 2010 by 5.6 percentage points. In other words, women fared better in the early 1990s compared to the early 2000s, but women made relative salary gains during the next decade. This result is similar to that of Blau and Kahn (2007). The initial OLS results indicate that the unadjusted salary gap for each sample declined when typical controls are included. Columns (2), (4), and (6) show salary gaps of 12.7%, 17.6%, and 13.5% for 1993, 2003, and 2010, respectively, once all controls are in the model. In particular, adding the occupation dummy variables reduces the coefficient on *female* in each of the regressions by about 10 percentage points. This finding is consistent with Goldin (2014), who used American Community Survey data to find a seven percentage point decline in the gap for college-educated, full-time workers once occupation was added.

Confident that our data produce results comparable to earlier research, we next consider how occupation matters to the salary gap by exploring occupational segregation, feminization, and family effects. As noted earlier, sociology and economics provide alternative explanations. Perhaps women are crowded out of higher-paying occupations that are somehow reserved for men. Hegewisch and Hartmann (2014) and Blau et al. (2012) suggest that even though women have made inroads into jobs typically identified as "male" occupations, occupational segregation is alive and well. Alternatively, as women move into higher paying occupations, the occupations are *feminized*, salaries decline for men and women (Mandel, 2013). We examine these possible explanations in more detail.[9]

Has Occupational Segregation Changed Over Time?

The term *occupational segregation* is used to describe a workforce in which identifiable groups defined by gender and/or race are situated in different jobs. The occupational segregation measurement has developed as researchers study the phenomenon. Blackburn (2012) suggests that the analysis should include a measure that can be decomposed into horizontal and vertical dimensions. Horizontal segregation refers to the level that men and women comprise different occupations. Vertical segregation assesses how much men and women are valued within similar job categories, usually by a measurement using salary by detailed occupation.

Following Blackburn and Jarman (1997), we employ the following geometric relationship:

$$O^2 = V^2 + H^2 \tag{2}$$

where overall segregation (O) is comprised of vertical (V) and horizontal (H) segregation components. Blackburn (2012) indicates that there are several measures of overall segregation, including the commonly employed dissimilarity index (DI). However, many of these indices cannot be used in the decomposition, so one cannot know if occupational segregation occurs/declines due to changes across occupations or within occupations. Blackburn suggests a Gini coefficient that estimates the inequality of each occupation via the gender ratio. The vertical segregation component is measured by an estimated Gini coefficient over average salary by occupation. Horizontal segregation is then deduced via Eq. (2).

Generally following Alonso-Villar, Del Rio and Gradin (2012), the Gini coefficient (G_f) measuring overall occupational segregation is derived by ordering the proportions of male and female workers, starting with the occupation with the lowest percentage of female workers. The measurement is:

$$G_f = \frac{\sum_{i,j} \frac{t_i t_j}{T\,T} \left| \frac{F_i}{t_i} - \frac{F_j}{t_j} \right|}{2 \frac{F_i}{T}} \tag{3.1}$$

Where there are J occupations , $\sum t_i/T$ represents the cumulative proportion of employment over all workers, F_i/T represents the cumulative proportion of females over all workers F_i/t_i is the cumulative proportion of females over all females , F_j/t_j is the proportion of females in occupation j, and T is the total number of workers.

The measure compares relative proportions by ordering the level of "femaleness" of the occupation from low proportions of females to high proportions of females. If all occupations contained the same proportion of females, G_f would be 0; if one occupation had all of the females in the workforce and no males, G_f would be 1.

The Gini coefficient for salary by occupation is:

$$G_y = \frac{\sum_{i=1}^{n} \sum_{j=1}^{n} |y_i - y_n|}{\frac{1}{n^2}} \tag{3.2}$$

where y_n is average salary for occupation j for group i, and n is the total number of occupations.[10]

In the NSCG data, there are over 120 occupations for each of the samples. The percent female and average salaries per occupation were calculated for each sample year to estimate the two Gini coefficients. The overall measure yields the same value as the Dissimilarity Index (Blackburn, 2012).[11] The overall measurements indicate that about 38%, 33% and 30% of men (women) would need to switch into a female (male) occupation in each of the sample years in order to eliminate occupational segregation. Thus, overall occupational segregation declined 5.1 percentage points from 1993 to 2003 and declined an additional 2.5 percentage points from 2003 to 2010.

Table 2 demonstrates the decline of horizontal segregation and increase of vertical segregation, suggesting that women are moving into traditionally male occupations and men increasingly populate traditionally female occupations. We observe increases in the salary differentials within occupations with men receiving larger salaries compared to women. A review of the salary ratio for the top ten occupations in which men dominate presents evidence of the changes over time (Table 3, next page). In some fields, including petroleum and mechanical engineering, as the percentage of women increase, the salary differential declines. This result is particularly true among occupations that are not typically recognized as requiring a college education (e.g. construction trades and mechanical repair). For electrical and mechanical technicians, the lowest percent female occupation in 2010, the salary gap was negative, indicating that women are paid about 7.5% more than their male counterparts as this type of technician. For physics professors, the percent female and the gap declined. The only occupation that showed an increase in the percent female and an increase in the salary gap is agricultural engineering.

Table 2
Occupation Segregation Decomposition

	1993	2003	2010
Overall Segregation	0.379	0.328	0.294
Vertical Segregation	0.125	0.130	0.171
Horizontal Segregation	0.358[a]	0.301	0.239
Number of Occupations	124	126	129

[a]Using Eq. (2), the horizontal segregation measurement is calculated as the square root of O2-V2. Estimates are rounded to three decimal places for presentation. Rounding may lead to slightly different values.

Table 3
Occupations with Lowest and Highest Female Percentage, Change from 1993 to 2010

Top Male Occupations in 2010					
	2010			1993	
Occupation	**Percent female**	**Differential**		**Percent female**	**Differential**
Electrical and mechanical technicians	7.006	-0.075		10.759	0.132
Agricultural engineers	8.333	1.337		2.439	0.590
Mechanical engineers	9.295	0.104		5.364	0.126
Engineering managers	9.406	0.060		n/a	n/a
Mechanics and repairers	9.479	-0.007		2.951	0.128
Computer network architect	9.677	-0.093		n/a	n/a
Construction trades	9.896	-0.275		3.436	0.107
Physics professors	10.145	0.075		14.545	0.283
Electrical and electronic engineers	10.594	0.110		5.320	0.116
Petroleum engineers	11.111	0.017		5.435	0.123
Surveyors	11.111	0.149		4.762	0.425
Top Female Occupations in 2010					
Health occupations, not classified	72.797	0.195		60.360	0.165
Counselors	74.084	0.078		63.295	0.060
Health technology and technicians	74.954	0.073		71.155	0.021
Social Workers	78.788	0.078		68.968	0.077
Librarians	79.221	0.247		73.992	0.080
Accounting clerks and bookkeepers	79.730	0.056		72.295	0.072
Special Education teachers	81.281	-0.024		82.334	0.037
Nurses, pharmacists, and dieticians	85.166	0.215		78.262	0.164
Elementary School teachers	85.205	0.132		85.444	0.083
Secretaries, receptionists and typists	92.151	0.032		94.872	0.030
Pre-K and Kindergarten teachers	94.709	0.255		95.590	0.000

Data Source: Authors' samples from the 1993 and 2010 NSCG.

Table 3 also presents the ten occupations with the highest percent female. In seven of the ten occupations, the percent female increased; of those seven, five exhibit an increase in the salary gap. Librarians now have a salary gap that is three times the size it was in 1993; only accounting clerks and bookkeepers have experienced an increase in the percent female and a decline in the gap over the time period. Special education remained female-dominated and experienced a decline in the salary gap, with women earning more than men in 2010. In the group of female-dominated occupations, Pre-K and Kindergarten teacher fared the worst; although the percentage of women declined slightly, the salary gap increased by nearly 26 percentage points.

The analysis suggests that women are gaining in historically male-dominated occupations, but women remaining in traditionally female occupations are experiencing salary declines compared to men. Thus, we see a decline in horizontal segregation but an uptick in vertical segregation. The decline in overall segregation seems to be dominated by the decline in horizontal segregation.

Is There a Feminization Effect?

As noted in the previous section, the salary gap has increased for a number of traditionally female-dominated occupations. K-12 teachers, librarians, and health professionals experienced both an increase in the feminization of the occupation and a larger salary gap. Sociologists interpret a negative relationship between the percent female of an occupation and salary as discrimination. Lower pay for female-dominated occupations may be due to "devaluing," in which work performed by women is valued less by employers than the work performed by men (Murphy and Oesch, 2016; Levanon et al., 2009; Mandel, 2013). To examine this possibility, we include the percent female as a control variable in the salary regression. Table 4 presents the coefficient estimates for *Female* and *Percent Female* variables.

Table 4
Feminization of Occupations, Regressions with Percent Female and Change in Percent Female

	1993		2003		2010	
Female	-0.114*** (0.003)	n/a	-0.159*** (0.007)	-0.182*** (0.008)	-0.102*** (0.001)	-0.101*** (0.009)
Percent Female	-0.002*** (0.0002)		-0.002*** (0.0002)		-0.003*** (0.0001)	
Change in Percent Female		n/a		-0.007*** (0.0007)		-0.004*** (0.0004)
Adj R^2	0.328		0.262	0.274	0.348	0.352
F-Statistic	1623.82***		437.48***	369.79***	258.89***	228.53***
N	106,772		52,667	40,420	47,664	45,378

Output from a regression model that includes controls for demographic, family, employer, and 15 occupations. Contact authors for detailed regressions.

***statistical significance at 1 percent.

**statistical significance at 5 percent.

*statistical significance at 10 percent.

The results indicate that the salary gap closes (compared to results in Table 1), suggesting at least 1 to 3 percentage points of the gap can be explained by occupational feminization. For each sample year, the coefficient on *Percent Female* is negative. The results suggest that a 1 percentage point increase in the percentage of women in an occupation leads to a 0.2% to 0.3% decline in salary. While this impact appears small, if the estimate is applied to the occupation with the highest level of feminization, pre-school and kindergarten teachers, the effect is approximately a 24% lower average salary compared to an occupation without any women. Is this evidence of a devaluing of predominately female occupations? The results are suggestive as many of these occupations have been predominately female for more than the seventeen years covered by this study, and the effect is consistent over the time period.

A second test examines whether a change in feminization affects the salary gap.[12] Table 4 presents the results for 2003 and 2010. The change in the percent female of an occupation is the differences between the current and previous decade's percent female for each occupation. Because some occupational codes changed between 1993 and 2003, there is a loss in the sample size. The change in sample size between 2003 and 2010 is due to the fact that a few additional occupations were added in 2010.

The coefficient on *Change in Percent Female* is negative for both sample years, which means that a one percentage point increase in the proportion of women in an occupation decreases average salary by 0.4% to .7%. If we use the average salary, $79,395, of the 2010 sample, each 1% increase in the proportion of women in an occupation reduces the average by $317.58.

Evidence of a Compensating Differential?

A compensating differential may explain the decline in the salary gap after controlling for occupation. The presence of a compensating differential may be interpreted to reveal that women forego salary when seeking employment that accommodates family obligations. Leonard and Stanley (2015) perform a meta regression analysis to find that a 9.4% marriage premium accrues to men. Using a fixed-effects model, the premium increases to 12.2%.

Regardless of the occupation, women still serve, on average, as the primary caregivers for children and spend more time performing household chores than male spouses (Parker, 2015).[13] In order to examine the effect of family-related preferences or constraints on the gender salary gap, we construct two interaction control variables in which *Female* interacts with *Marital Status* and *Number of Children*. While imperfect, the interaction terms allow us to compare married women and men with children to unmarried men. If the coefficient estimate for both men and women with children is negative, we will have evidence that workers forego salary to meet family-related preferences or constraints.

The results, presented in Table 5, suggest that while males experience a positive salary effect with marital status and children, females experience a negative effect. The effect is relatively consistent across samples for marital status, suggesting that married women earn 9% to 11% less than men who are unmarried, while married men earn 10% to 14% more than unmarried men. Men with children experience a premium for each child, while women in 1993 and 2003 receive approximately 2% less salary with each child.

Table 5
Compensating Differential Regressions with Interaction Terms

	1993	2003	2010
Female	-0.034** (0.005)	-0.080** (0.018)	-0.031*** (0.018)
Married	0.103*** (0.004)	0.142*** (0.012)	0.142*** (0.014)
Kids	0.018*** (0.002)	0.036*** (0.004)	0.031*** (0.004)
Female*Married	-0.087*** (0.006)	-0.107*** (0.017)	-0.088*** (0.021)
Female*Kids	-0.027*** (0.003)	-0.028*** (0.006)	-0.008 (0.008)
Adj R^2	0.331	0.276	0.354
F-Statistic	1557.40***	357.58***	254.86***
N	106,772	40,420	45,378

Output from a regression model that includes controls for demographic, family, employer, and 15 occupations. Contact authors for detailed regressions.

***statistical significance at 1 percent.

**statistical significance at 5 percent.

*statistical significance at 10 percent.

The coefficient for 2010 indicates that women are unaffected by children. After adding the interaction terms to the model, the coefficient estimate on *Female* is lowered, suggesting that unmarried women without children earn less than their unmarried counterparts, but the difference is only a 3% gap by 2010. This result indicates that at least some of the gender salary gap can be explained by the effect of marital status on women in 2010, and for women with children in the earlier decades.

We do not know whether women accept lower pay to be home or if men and women are treated differently due to family obligations as a form of discrimination. If married men with children also experienced lower salary compared to unmarried childless men, then the compensating differential explanation would offer more support; however, the opposing effects by gender suggest the possibility of different treatment in the workplace. It appears that women are paid less while men receive a premium when married with children.

Discussion of Results

The three surveys at different time periods cover a nearly thirty-year span for analytical use. Our findings show less overall segregation as men and women move into occupations in which the other represents the majority. This decline corresponds to a decline in horizontal segregation as outlined by Blackburn (2012). Vertical segregation, the segregation within an occupation, increases slightly. The analysis reveals that women moving into male-majority occupations experience gains in salary, while women who remain in traditionally female-dominated occupations experience salary declines. A traditional economic explanation posits that the output of these occupations is less valued by the marketplace independent of the sex of the worker. One explanation suggests that market conditions have driven down market returns for the occupation. In this scenario, the supply of workers has increased faster than the demand, imposing downward pressure on the wage. A second potential explanation considers the structure of the occupations. Might it be that female-dominated occupations accommodate marriage and family constraints or preferences? Unfortunately, our data do not permit us to examine the structure of employment arrangements or assess the motivation for accepting particular employment, so we cannot answer this question and must leave it for future research.

With respect to the increase in vertical integration, while pharmacists may be an exception, we are aware that labels like "doctor" and "lawyer" and "accountant" do not represent homogenous occupations. Surgeons and general practitioners are doctors as are podiatrists and optometrists, but salaries are not similar. If men systematically gravitate toward higher paying specialties within an occupation or women systematically gravitate toward lower paying specialties within an occupation, we will observe vertical segregation for reasons relating to the specialization within an occupation rather than sex discrimination. Regardless, the more interesting question is whether women are systematically discouraged or deterred from employment in higher paying specialties or whether the observed segmentation reflects preferences that differ systematically by gender. It would be very interesting for future research to examine individual occupational titles within broader categories to determine how gender informs variation in type of employment and earnings. While we expect male and female thoracic surgeons to earn a very similar salary, controlling for the typical variables, why are there so few women employed as thoracic surgeons relative to men?

Goldin and Katz (2014) contend that highly educated women are likely selecting jobs within occupations that pay less in exchange for "temporal flexibility." Two law school graduates from the same institution with identical transcripts might very well pursue distinctly different careers. It may be that the man pursues employment in a private firm that demands 60 hours per week, travel, and work on nights and weekends, while the women pursues a job that permits greater "temporal flexibility" such that women may work the same number of hours, but she determines when the hours are worked. While both are lawyers, the male receives a premium to offset the lack of "temporal flexibility." In effect, he is paid to work where and when he is told to work. Goldin and Katz (2011) address the cost of workplace flexibility among highly educated workers, specifically highly-educated professionals. They find that the penalties have declined in many professions, and they expect endogenous and exogenous factors are likely to contribute to further declines across more occupations.

An interesting finding regarding Pre-K and Kindergarten teachers reveals a 26% premium accruing to men. While the NSCG data indicate that slightly more men comprise the occupation in 2010 compared to 1993, the discrepancy motivates questions regarding why men are paid so much more. A review of our 2010 sample suggests several reasons for the large pay gap. While there are very few men in the Pre-K and

Kindergarten occupation (about 5%), all the men are in public sector education. In contrast, approximately 6% of the women are self-employed or in for-profit organizations, likely daycare centers, that typically pay much lower than public school systems. Further, about 45% of the women work for employers with less than 100 employees, compared to about 11% of the men. The men are also located in regions where teacher unions are prevalent. Women are dispersed across the country with 37% in the South, where teacher salary is typically lower compared to other regions.[14] While 39% of women in this occupational category have graduate degrees compared 11% of men, the type of employer and location variables lead to a larger gender gap in this occupation.[15]

The analysis includes two measures of occupational feminization and indicates that as the percentage of feminization increases, the salary for individuals within that occupation decreases. This result is consistent with the devaluation theory of social scientists. Levanon et al. (2009) examine fifty years of U.S. Census data and find a similar result. The feminization effect occurs for both genders; as women enter into male occupations, wages decline for both men and women. In our study, we find a smaller effect due to feminization of 0.2% to 0.3%; in contrast Levanon et al. found percentages as high as 10% for some decades. In addition, our results demonstrate that the coefficient on *female* declines by 1 to 4 percentage points, indicating that feminization plays a role in the salary gap that the binary control for gender subsumes when entered into the model alone.

Our analysis also indicates that married women with children earn less than married men with children and all other groups. This finding is consistent with laboratory and field experiments results of Correll, Benard, and Paik (2007), who find that mothers face discrimination compared to nonparents. The compensating differential hypothesis contends that workers forego earnings to accommodate obligations and preferences associated with marriage and family. Evidence supporting the theory would have to show that married men with children report a tradeoff compared to single men. To the contrary, our findings reveal that married men earn more than single men as well as and married women, thus offering little support for the compensating differential hypothesis. How then can we explain that married men earn more? There are some rational possibilities to explain these findings.

Firms may discount single men relative to married men because single men, who have fewer responsibilities to others, may be more likely to seek employment elsewhere. Turnover is a cost to firms that may be built into the structure of wages. This is an empirical question that our data cannot address. Nonetheless, one is likely to concede that it is "easier" to change jobs when one has only oneself to consider. In contrast, married workers must consider the interests and preferences of spouses and children. There is a reason that the society refers to marriage and family as "settling down."

With regard to the lower average salary for married women, especially married women with children, several other studies have found a similar result and suggest that children result in lower productivity for primary caregivers or that workplace discrimination still exists for this group. An international survey conducted by The Economist and YouGov indicates that women are still the primary caregivers in society and they scale back on work hours once children are added to the family (Economist, 2017). Budig and England (2001) study the motherhood penalty with ten years of data from the National Longitudinal Survey of Youth and find that the temporal flexibility theory posited by Goldin and Katz (2016) weakly explains the effect on married women. Instead, they suggest that the motherhood penalty results from lost productivity due to children and workplace discrimination. Women interviewed by Ali (2016) for an NBC article suggest that mothers tend to face a higher standard because they are assumed less committed to the work. As long as women are assumed to be the primary caregivers in society, that role, while external to the workplace, imbues employer attitudes about what and how women can do.

Conclusion

Men continue to receive higher salary than women after controlling for a host of occupational, human capital and demographic variables. The gender salary gap for college educated individuals has remained relatively constant between 1993 and 2010. A good proportion of the gap can be explained by occupation and occupational segregation. Salaries rose for women entering traditional male-dominated occupations; for the women remaining in occupations dominated by women, salary declined. The decrease in overall occupational segregation is driven by declines in horizontal segregation as women enter traditional male occupations and men enter occupations that are disproportionately female. While there is evidence that occupational feminization affects the salary gap, which suggests the devaluation theory predominates, we also cannot control for market outcomes for occupations, and therefore leave the final analysis to future research using data suited to the purpose.

We find that men receive a premium for marriage and children while women experience a discount. To reveal a compensating differential, the analysis needs to demonstrate that both men and women incurred a discount to accommodate family obligations. That there is a small salary gap for single women without children, which magnifies the discount imposed on married women with children. Highly educated married women with children may face workplace discrimination; as the primary caregiver, they may face productivity losses, or they maximize utility by trading income for "temporal flexibility." If the latter is true, the exchange of earnings for time represents social progress as women obtain influence (power or control) to dictate the terms of their employment.

The absolute gains in salary by women reflect an expansion of educational and employment opportunities. Closing the gender pay gap represents gains relative to men. The persistence of a gender wage gap represents an injustice if women are in any way denied employment opportunities afforded to men or are paid less than men to perform identical work. The documented case of pharmacists suggests that whatever gender discrimination remains embedded in society does not always interfere with market forces that determine earnings, especially in high salary occupations. Salary gains by women who enter male-dominated occupations offer further evidence that market forces operate as expected when women and men have comparable opportunities for education, training, and employment.

Overall occupational segregation estimates do not account for the various types of employment within an occupation. Progress toward understanding the structure, causes, and implications of gender salary inequality will likely be achieved by examining the vertical segregation within occupations in greater detail than we can do here. Several factors seem to affect the gender pay gap, not all of which are purely economic in nature. Ultimately, understanding the gender pay gap will likely require that economic analyses incorporate the insights gained from related social science disciplines such as sociology. Once a comprehensive understanding of the phenomenon is established, society can correct the attitudes and practices that have denied women opportunities and income disproportionately accruing to men throughout history despite political and legal initiatives to affect change. The gender gap will close, either literally and figuratively, once the latent discrimination found in institutions and society are revealed and transparent.

Acknowledgments

The authors thank Xiaoyan (Laura) Youderian for her discussant comments at the 2017 Midwest Economics Association meetings and Amanda Cook of Bowling Green State University for her helpful comments.

Endnotes

[1] One could add a fourth explanation: Babcock and Lashever (2003) contend that the pay gap reflects the superior negotiation skills of men. We are unable to examine negotiation skill as part of this study as the data are unavailable.

[2] Levanon et al. (2009) indicate that these theories often overlap and cannot explain all of the wage gap. Generally, sociologists acknowledge the importance of human capital development (education, training) and other factors that determine wages and therefore include controls for these factors in their models.

[3] The NSF conducted the NSCG survey in 2010 as well as 2013. We opted to employ the 2010 data because it was available when we began our research.

[4] The first two samples were derived from the decennial census (finding the size of the population and sending mailings to respondents). The 2010 sample is derived from the American Community Survey and the 2008 National Survey of Recent College Graduates (NSRCG). The change was due to cost efficiency

decisions as the size of the college graduate and science population increased. The NSCG will ultimately yield smaller survey samples in the future (Sf-83-1 Supporting Statement).

[5] The authors explored the possibility of occupational endogeneity with a selection model and a GMM model using IV estimation. In the selection model, a F-test indicated that the Mills ratios jointly contributed to the model, but generally, the coefficients were small or individually statistically insignificant, suggesting an overall negligible selection effect. The GMM model yielded coefficients on *Female* that were not consistent with those from the OLS model. We suspect that the results were unwieldy due to many of the first-stage estimated probabilities being very small for many of the observations. Because we were unsure of the validity of these results, we opted for the conventional OLS model.

[6] The NSCG does not include industry identifiers. The occupational and private/public variables help to control for some of the industry effect (e.g., law, university professors, social workers) To the degree that some industries have larger gender pay gaps than others, then the occupational and public/private coefficients may be larger than they would otherwise be; however, since the paper focuses on the independent effect on the gender variable, the result should not be particularly affected.

[7] The means and standard deviations of the samples are available from the authors on request.

[8] The full regressions are available from the authors.

[9] Sociologists pose a third theory, queueing. Queueing theory suggests that pay rates affect the proportion of females in an occupation. However, a number of sociologists have not found a reverse causality between wages and female proportion (Levanon, England, and Allison 2009; England, Allison, and Wu 2007; Semyonov and Lewin-Epstein, 1989). We therefore elect to focus on remaining theories in social science theory.

[10] Blackburn (2012) states that one orders the lowest paying to highest paying occupations to create the vertical dimension.

[11] Dissimilarity indices are available from the authors on request.

[12] Karlin et al. (2002) find that percent female and change in percent female are not endogenous to the salary model.

[13] The NSCG does not distinguish couples based on sexual orientation. We only know if the respondent is married or not, and the respondent's gender, not the spouse's gender.

[14] The National Education Association reports that the average teacher salary for all southern states is below the national average and all "southern" states except Georgia (24/51) are listed in lower half of the ordinal ranking of the fifty states and the District of Columbia.

[15] Results available from the authors on request.

References

Ali, Safia Sami, 2016. "'Motherhood Penalty' Can Affect Women Who Never Even Have a Child." Published at *NBC News*. https://www.nbcnews.com/business/careers/motherhood-penalty-can-affect-women-who-never-even-have-child-n548511. [December 8, 2017].

Alonso-Villar, Olga, Coral Del Rio, and Carlos Gradin, 2012. "The Extent of Occupational Segregation in the United States: Differences by Race, Ethnicity, and Gender." *Industrial Relations*, Vol. 51, No. 2, pp.179-212.

Ashenfelter, Orley and Cecelia E. Rouse. "Income, Schooling and Ability: Evidence from a New Sample of Identical Twins." *Quarterly Journal of Economics* Vol. 113, No.1, pp. 253-284.

Babcock, Linda and Sara Lashever, 2003. *Women Don't Ask: Negotiation and the Gender Divide.* Princeton University Press.

Baxter, Emily, 2015. "How the Gender Wage Gap Differs by Occupation. Center for American Progress Organization." Posted April 14, 2015. https://www.americanprogress.org/issues/women/news/2015/04/14/110959/how-the-gender-wage-gap-differs-by-occupation/ [April 30, 2015].

Benedict, Mary Ellen, David McClough and John Hoag, 2012. "STEM: A Path to Self-Employment and Jobs." *Journal of Entrepreneurship Education* 15, pp. 99-122.

Björklund, A., M. Lindahl, and E. Plug, 2006. "The origins of intergenerational associations: Lessons from swedish adoption data." *The Quarterly Journal of Economics* Vol. 121, No. 3, pp. 999–1028.

Blackburn, Robert M. and Jennifer Jarman, 1997. "Occupational Gender Segregation." *Social Research Update*, 16. http://sru.soc.survey,ac.uk/SRU16/SRU16.html. [January 16, 2016]

Blackburn, Robert M., 2012. "The measurement of occupational segregation and its component dimensions." *International Journal of Social Research Metholodology*, Vol. 15, No. 3, pp. 175-198.

Blau, Francine D. and Wallace E. Hendricks, 1979. "Occupational Segregation by Sex: Trends and Prospects." *Journal of Human Resources*, Vol. 14, No. 2, pp.197-210.

Blau, Francine D. and Lawrence M. Kahn, 2007. "The Gender Pay Gap: Have Women Gone as Far as They Can?" *Academy of Management Perspectives*, Vol. 21, No. 1, pp. 7-23.

Blau, Francine D., Peter Brummund, and Albert Yung-Hsu Liu, 2012. "Trends in Occupational Segregation by Gender 1970-2009: Adjusting for the Impact of Changes in the Occupational Coding System." Natural Bureau of Economic Research Working Paper 17993. http://www.nber.org/papers/w17993 [January 2016]

Borjas, George, 1996. *Labor Economics*. New York: McGraw-Hill.

Budig, Michelle J. and Paula England, 2001. "The Wage Penalty for Motherhood." *American Sociological Review*, Vol. 66, No. 2, pp. 204-225.

Case, Ann and Angus Deaton, 2017. "Mortality and morbidity in the 21st century." Brookings Papers on Economic Activity, Washington, DC: Brookings Institute, pp. 397-476.

Correll, Shelley J., Stephen Benard, and In Paik, 2007. "Getting a job is there a motherhood penalty?" *American Journal of Sociology*, Vol. 112, No.5, pp. 1297-1338.

Cotter, David A., Joann M. DeFiore, Joan M. Hermsen, Brenda Marstellar Kowalewski, and Reeve Vannerman, 1995. "Occupational Gender Desegregation in the 1980s." *Work and Occupations*, Vol. 22, No. 3, pp. 3-21.

Dubner, Stephen J., 2016. "The True Story of the Gender Pay Gap." *Freakonomics*, Retrieved from http://freakonomics.com/podcast/the-true-story-of-the-gender-pay-gap-a-new-freakonomics-radio-podcast [January 21, 2016]

Ebenharter, Veronike, 2006. "Family Background, Occupational Choice and Intergenerational Income Mobility - Germany and the United States compared." Unpublished paper found at http://www.diw.de/documents/dokumentenarchiv/17/44460/eberharter.pdf. [March 15, 2017]

England, Paula, 1981. "Assessing Trends in Occupational segregation, 1900-1976." In Ivar Berg ed., *Sociological Perspectives on Labor Markets*, New York, NY: Academic Press, pp. 273-95.

England, Paula, Paul Allison, and Yuxiao Wu, 2007. "Does Bad Pay Cause Occupations to Feminize, Does Feminization Reduce Pay, and How Can We Tell with Longitudinal Data?" *Social Science Research,* Vol. 36, No. 3, pp. 1237-56,

Gill, Andrew M. and Duane E. Leigh, 2000. "Community College Enrollment, College Major and the Gender Wage Gap." *Industrial and Labor Relations Review*, Vol. 54, No. 1, pp. 163-181.

Goldin, Claudia and Lawrence F. Katz, 2002. "The Power of the Pill: Oral Contraceptives and Women's Career and Marriage Decisions." *Journal of Political Economy*, Vol. 110, No. 4, pp. 730-770.

Goldin, Claudia and Lawrence F. Katz, 2011. "The Cost of Workplace Flexibility for High-Powered Professionals." *Annals of the Academy of Political and Social Science*, Vol. 638, No. 1, pp. 45-67.

Goldin, Claudia, 2014. "The Grand Gender Convergence in its Last Chapter." *American Economic Review*, Vol. 104, No.4, pp. 1091-1119.

Goldin, Claudia and Lawrence F. Katz, 2016. "A Most Egalitarian Profession: Pharmacy and the Evolution of a Family-Friendly Occupation." *Journal of Labor Economics*, Vol. 34, No. 3, pp. 705-746.

Heckman, J., 1979. "Sample selection bias as a specification error." *Econometrica*, Vol. 47, No. 1, pp. 153-161.

Hegewisch, Ariane and Heidi Hartmann, 2014. "Occupational Segregation and the Gender Wage Gap: A Job Half Done." Institute of Women's Policy Research. https://iwpr.org/publications/occupational-segregation-and-the-gender-wage-gap-a-job-half-done. [February 10, 2016]

Karlin, Carolyn Aman, Paula England, and Mary Richardson, 2002. "Why Do 'Women's Jobs' Have Low Pay for Their Educational Level?" *Gender Issues*, 20(4), pp. 3-22.

Leonard, Megan de Linde and T.D. Stanley, 2015. "Married with children: What remains when observable biases are removed from the reported male marriage wage premium?" *Labour Economics*, 33, pp. 72-90.

Levanon, Asaf, Paula England, and Paul Allison, 2009. "Occupational Feminization and Pay: Assessing Causal Dynamics Using 1950-2000 U.S. Census Data." *Social Forces*, Vol. 88, No. 2, pp. 865-892.

Loury, Linda Datcher, 1997. "The Gender Earnings Gap among College-Educated Workers." *Industrial and Labor Relations Review*, Vol. 50, No. 4, pp. 580-593.

Mandel, Hadas, 2013. "Up the Down Staircase: Women's Upward Mobility and the Wage Penalty for Occupational Feminization, 1970-2007." *Social Forces*, Vol. 91, No.4, pp. 1183-1207.

Mathews, T. J. and Brady E. Hamilton, 2009. "Delayed childbearing: More women are having first child later in life." National Center for Health Statistics Data Brief, no. 21. Center for Disease Control. https://www.cdc.gov/nchs/data/databriefs/db21.pdf. [November 21, 2017].

Mincer, Jacob and Solomon Polachek, 1974. "Family Investments in Human Capital: Earnings of Women." *Journal of Political Economy*, Vol. 82, No. 2, pp. S76-S108.

McClough, David and Mary Ellen Benedict, 2017. "Not All Education Is Created Equal." *American Economist*, Vol. 62, No. 2, pp. 184-205.

Murphy, Emily and Daniel Oesch, 2016. "The Feminization of Occupations and Change in Wages: A Panel Analysis of Britain, Germany, and Switzerland." *Social Forces*, Vol. 94, No. 3, pp. 1221-1255.

National Center for Education Statistics (2017). Table 104.20. Percentage of persons 25 to 29 years old with selected levels of educational attainment, by race/ethnicity and sex: Selected years, 1920 through 2016. Institute for Education Science. https://nces.ed.gov/programs/digest/d16/tables/dt16_104.20.asp. [November, 2017].

Oaxaca, Ronald, 1973. "Male-Female Wage Differentials in Urban Labor Markets." *International Economic Review*, Vol. 14, No. 3, pp. 693-709.

Parker, Kim, 2015. "Women more than men adjust their careers for family life, FactTank News in the Numbers." Pew Research Center. http://www.pewresearch.org/fact-tank/2015/10/01/women-more-than-men-adjust-their-careers-for-family-life. [January 25, 2018].

Petersen, Trond, and Ishak Saporta (2004). The Opportunity Structure for Discrimination. *American Journal of Sociology*, Vol. 109, No. 4, 852-901.

"Rankings of the States 2016 and Estimates of School Statistics 2017" Washington, DC: National Education Association. http://www.nea.org/assets/docs/2017_Rankings_and_Estimates_Report-FINAL-SECURED.pdf. [March 10, 2017].

Salas-Velasco, Manuel, 2010. "Wage determinants among medical doctors and nurses in Spain." *Higher Education*, Vol. 60, No. 4 pp. 357-368.

Schneider, Jennifer and Elise Gold, 2016. "Women's work' and the gender pay gap." Economic Policy Institute. https://www.epi.org/publication/womens-work-and-the-gender-pay-gap-how-discrimination-societal-norms-and-other-forces-affect-womens-occupational-choices-and-their-pay. [September 19, 2016]

Semyonov, Moshe, and Noah Lewing-Epstein, 1989. "Segregation and Competition in Occupational Labor Markets." *Social Forces*, Vol. 68, No. 2, pp. 379-96.

SF-83-1 SUPPORTING STATEMENT for 2010 National Survey of College Graduates, https://www.reginfo.gov/public/do/DownloadDocument?objectID=17891902. [March 3, 2017].

Solberg, Eric, 2005. "The Gender Pay Gap by Occupation: A Test of the Crowding Hypothesis." *Contemporary Economic Policy*, Vol. 23, No. 1, pp. 129-148.

"The gender pay gap." *The Economist*, October 7, 2017. https://www.economist.com/news/international/21729993-women-still-earn-lot-less-men-despite-decades-equal-pay-laws-why-gender. [December 3, 2017]

U.S. Department of Education, National Center for Education Statistics, Integrated Postsecondary Education Data System (IPEDS), Fall 2012, Completions component.

Vella, Francis, 1993. "Gender Roles, Occupational Choice and Gender Wage Differential." The Economic Record, Vol. 69, pp. 382-392.

IV. Changing Union Strategies: American, European, Transatlantic

German Trade Union Attempts to Export Power

STEPHEN J. SILVIA
American University

Abstract

Decades of investment abroad by German transnational enterprises has created global supply chains where costs are lower and employee representation weaker. Production abroad has exposed German managers to far more confrontational employment relations regimes than the social partnership practiced at home. These trends have increased German firms' leverage over domestic unions. To meet these challenges, German trade unionists have experimented with exporting power. This contribution investigates five cases in the metalworking and communications sectors of investment in the European Union and the United States. The research shows that unions still have an extraordinarily difficult time exporting power beyond

Introduction

Postwar German trade unionists have always engaged with other labor movements beyond their borders, but the issues, tactics and strategies have changed substantially over the decades as the German economy, the global economy, technology and the capacities of German trade unions have changed. Immediately after the Second World War, the recovering independent labor movement in what was soon to become West Germany[1] accepted support from allied trade unions, which in turn received funds for this purpose from allied governments. The foreign policy of West German trade unionists initially focused on two somewhat incongruous goals: German unity and isolating the East German labor movement because of its leadership's subservience to the dominant pro-Soviet Socialist Unity Party of Germany (Slobodian 2010).

Starting in the 1960s, the blossoming of the West German "economic miracle" enabled a more self-assured and financially secure West German trade union movement to grow in influence internationally and to take positions that differed from its former foreign trade union patrons, in particular, those from United States. These independent stances included the cultivation of positive relations with trade unions in the Soviet bloc, South Africa and Latin America to promote *détente*, combat apartheid and condemn military juntas (Fichter 2010). Despite the West German labor movement's growth in confidence and autonomy in international affairs between the 1960s and 1980s, policy in this area remained largely an echo chamber of the high politics of the day combined with assistance principally in the form of cash and training in the basics of trade union work to ideologically sympathetic unions in less affluent countries.

The end of the Cold War at the start of the 1990s brought capitalism and democracy to most of central and eastern Europe, which opened things up for both foreign direct investment (FDI) and the formation of autonomous trade unions. When Germany began to run regular massive current account surpluses in the mid 2000s, German firms used them to finance the creation and expansion of global value chains (GVCs). The emergence of German-led GVCs and the recent proliferation of European Union efforts to form free trade areas sparked concern among leaders of German employee organizations (i.e., trade unions and works councils) that they could soon lose their relatively secure position in their domestic labor market. Union leaders responded by increasing their organizations' engagement with weaker trade union movements abroad

Author's address: American University, 4400 Massachusetts Ave., NW, Washington, DC 20016-8071 where German firms have substantial investments to help improve the capacity of workers at those investment sites to organize and to bargain collectively (Hübner 2015: 3-7).

This contribution assesses whether the recent surge in assistance from German employee organizations has affected employment relations in other countries or, short of that, at least substantially changed the course of events. The article reports on five cases: (1) attempts involving the German mechanical engineering employees union, IG Metall, to coordinate collective bargaining in transnational European regions; (2) IG Metall and works councils efforts to assist the United Auto Workers (UAW) in organizing drives at German automobile producers and parts suppliers in the United States; (3) IG Metall's undertakings to strengthen a Hungarian mechanical engineering union, VASAS; (4) the work by IG Metall and the Siemens enterprise works council to obtain a neutrality agreement for US unions interested in organizing Siemens USA facilities; and (5) cooperation between the German service employees union, ver.di, and the Communications workers of America (CWA) to organize Deutsche Telekom's subsidiary, T-Mobile USA.

The research shows that despite the information technology revolution, which arguably has reduced the barriers to collective action, exporting trade union power across borders remains difficult. The laws of other countries often restrict options for assistance, the residual reluctance of German employee representatives to use political capital and funds on causes that do not directly benefit their members, and hostile political dynamics abroad readily impede German employee representatives' attempts to engage in transnational assistance. The cases show that taking a strategic approach as well as dedicating significant financial and personnel resources are necessary to making a difference, but by no means guarantee success.

German Labor and European Transnational Collective Bargaining

The early 1990s were a time of great change in Europe. The Cold War had just ended, the member states completed the program to create a single European market and the Maastricht Treaty on European Union came into force. In March 1993, the leadership of the European Metalworkers Federation (EMF), which was a confederation of European trade unions that organized employees in the mechanical engineering sector, held its first collective bargaining conference in order to develop policies for a much more integrated European economy. The conference produced a statement of principle asserting that wage settlements throughout Europe should equal the sum of two components: the inflation rate "to protect real wages" plus "a share in productivity increases for employees," to preserve labor's portion of national income (Dufresne 2002: 80-81). Unions in some countries and sectors had used this combination of inflation and productivity growth since the 1950s to estimate the "wage space" available in collective bargaining (Ehrenberg 2008), but it had not been embraced as a European norm. Two years later, the eighth EMF general assembly set "increased coordination at the European level" as a goal in the organization's 1995-1999 action program, "including prior consultations on bargaining demands and the possible presence of invited observers from other EMF affiliated unions before and/or during the bargaining process" (EMF 1995).

The motivations behind the EMF actions were not only to advance transnational collective bargaining, but also to defend against a trend toward "competition oriented" collective bargaining, which stressed reducing unit labor costs (Schulten 1998: 483-84). A prominent example of the competition-oriented approach is the 1996 Belgian "Law for Promotion of Employment and Preventive Safeguarding of Competitiveness." The law requires Belgium's Central Economic Council to enforce a ceiling on wage settlements; they may not exceed a "wage standard," which is calculated using compensation trends from Belgium's three largest trade partners: France, Germany and the Netherlands (Vis and Woldendorp 2015: 161).

The 1996 law prompted Belgian union officials in the metalworking sector to invite their counterparts from Germany and the Netherlands to meet in June 1997. At the meeting, they agreed to create a transnational working group to exchange information regularly on collective bargaining developments and to hold an annual summit meeting. IG Metall quickly became the backbone for these transnational collective bargaining policy networks in the European metalworking sector. IG Metall districts formed regional networks with the metalworkers unions of neighboring countries and pushed for a reciprocal right of union officials from neighboring countries to participate as observers in collective bargaining talks. In the second half of 1997, Belgium and Dutch union officials participated as observers in the collective negotiations for the

German steel industry, which IG Metall's North Rhine-Westphalia district conducted. IG Metall Baden-Württemberg built a network with Italian and Swiss unions; as did IG Metall Bavaria with Austrian, Croatian, Hungarian, Slovakian and Slovenian unions; IG Metall Berlin-Brandenburg with Czech and Polish unions; IG Metall *Küste* (Coast) with Nordic trade unions; IG Metall Hesse with French unions; and IG Metall Lower-Saxony with the Amicus trade union in Great Britain (Gollbach and Schulten 2000: 166-173).

Union officials from Belgium, Germany and the Netherlands held their first summit in the Dutch town of Doorn in September 1998. Participation expanded beyond the metalworking sector to include officials from the peak confederations of the respective countries, as well as unions from the chemicals, construction, and private and public service sectors. The summit set auspicious objectives. The participants adopted a joint policy statement called the Doorn Declaration that embraced compensation increases equivalent to "the sum total of the evolution of prices and the increase in labor productivity" as a common bargaining objective. The Doorn Declaration also stated that participating unions should strive for "not only a fortification of mass purchasing power, but also job-creating measures (e.g., working time reduction)." The participants called for greater training opportunities for the employed and unemployed, and committed to "inform and consult one another regularly regarding the development of collective bargaining policy" through "a coordinating group of experts, who regularly meet in order to exchange information and experiences regarding collective bargaining negotiations." They also agreed "to examine how they can back up their demands beyond national frontiers when necessary" (https://www.boeckler.de/pdf/ta_doorn_erklaerung_1998.pdf: 2-3).

The Doorn Initiative is illustrative of both the promise and limitations of transnational union cooperation. The metalworkers unions in Belgium, the Netherlands and North Rhine-Westphalia exchanged observers for the 1998-99 collective bargaining round in the mechanical engineering sector. Observers participated in the preparation and assessment of negotiations. The German employers association let foreign observers sit in on the negotiations, but Belgian and Dutch employers did not. The EMF's 2005 collective bargaining conference included an evaluation of efforts to coordinate collective bargaining had begun a half a decade earlier. The evaluation stressed that employers fiercely resisted accepting inflation plus productivity growth as a reference point. Consequently, few unions were able to achieve that target. The evaluators also found that incompatibilities across national bargaining systems made coordination difficult, and language differences and limited resources also hampered the effort. The onset of the 2008-09 global financial crisis and the euro crisis dealt knockout blows to both regular regional summits of trade union leaders and observers from other unions in collective bargaining talks (Pulignano 2010; Gollbach and Schulten 2002, 175; Schulten 2002: 24).

The only institutional innovation from these efforts that has survived is collective bargaining information networks. In November 1999, the EMF established the European Collective Bargaining Information Network (EUCOB@). EUCOB@ was a network of "correspondents" from each EMF affiliated union who submit regular reports on collective bargaining developments. In 2003, the European Public Service Union, which is the European trade union federation for the public sector, inspired by the EMF, created its own collective bargaining network called EPSUCOB@ (Pond 2009: 315). In 2012, the EMF merged with its counterparts in the chemicals and textiles sectors to form IndustriALL Europe. IndustriALL Europe expanded EUCOB@ to include the chemicals and textiles sectors and renamed it EUCOB@N. Today, EUCOB@N has 175 active correspondents, holds regular meetings and frequently releases reports on collective bargaining developments in Europe (https://news.industriall-europe.eu/p/eucoban).

In summary, the Doorn process was never able to live up to the aims of its founders, that is, to add a transnational dimension to collective bargaining. To be sure, transnational collective bargaining policy information networks, which have survived, have value because they make more transparent compensation and working conditions throughout Europe, but there is no credible claim that these networks have led to better collective bargaining agreements or even influenced the course of events (Höpner and Seeliger 2017). As a result, German union leaders have been experimenting with more direct actions to diminish the threat of low-cost competition from foreign subsidiaries of German firms.

IG Metall, the UAW and Attempts to Organize German Auto Producers in the USA

The relationship between IG Metall and the UAW has changed substantially over the years. From immediately after the Second World War through the 1960s, the UAW was the leading partner and high politics dominated the relationship. The paths of the two unions diverged thereafter. A series of sales downturns in the US automobile industry starting in the mid 1970s and a shift in automobile production to the southern United States, which has traditionally been hostile to organized labor, reduced the power and influence of the American automobile employees union. Making matters worse, the UAW leadership became increasingly insular and embraced protectionism as the answer to the US auto industry's competitive problems. IG Metall, in contrast, remained the dominant German trade union, and influential in the German economy and society owing to continued German strength in the mechanical engineering sector, a much more supportive set of industrial relations institutions and generally competent leadership that took a far more strategic and holistic approach to formulating policy than its American counterpart. By the 1980s, IG Metall had become the stronger of the two unions and the relationship between the two unions had become "quite dissonant" (Fichter 2017: 2).

German foreign direct investment in the US automobile sector has an uneven record. The first attempt ended in failure. In 1978, Volkswagen bought an unfinished Chrysler plant in Westmoreland County, Pennsylvania, to produce the Rabbit (a.k.a. Golf). IG Metall officials and VW's works councils were supportive of the investment because it did not pose an immediate threat to employment or wages in Germany. Volkswagen management, at the urging of the employee representatives on the company's supervisory board, recognized the United Automobile Workers from the start. Delays in model upgrades and habitual labor strife left Volkswagen vulnerable to Japanese competition. The company increasingly lost market share and it ultimately closed the plant in 1988 (Silvia 2016: 32-34).

In the first half of the 1980s, the two high-end German producers – *Bayerische Motoren Werke* (BMW) and Daimler – greatly expanded sales in the United States on the back of a sharp appreciation of the United States dollar. In the second half of the decade, a substantial decline in the dollar and a tax reform ending deductions of interest payments on auto loans cut deeply into the US sales of the two German luxury automobile producers. BMW and Daimler management's response was to invest in the United States. BMW opened a plant in Spartanburg, South Carolina, in 1994. Daimler's facility started operating in Vance, Mississippi, in 1997. Unlike Volkswagen's Westmoreland county plant, the Spartanburg and Vance plants have been economic successes. In each case, the company initially produced a new model intended primarily for the US market. As a result, IG Metall officials and the works councilors for both companies backed the initial investment based on assurances from management that it did not threaten jobs or wages in Germany. A second difference between Volkswagen Westmoreland County and the other two plants is that they are in the South and neither BMW nor Daimler has recognized the UAW (Silvia 2016: 9-10 and 14-15).

The third chapter of German foreign direct investment in the United States automobile sector ended as a spectacular failure, but it facilitated a rapprochement between IG Metall and the UAW. In 1998, Daimler-Benz paid $37 billion in stock swaps for the Chrysler Corporation. DaimlerChrysler was legally incorporated as a German firm, which meant that it was subject to the German 1976 codetermination act requiring half of the twenty-member supervisory board to be employee representatives (Silvia 2013: 53). Chrysler also had a history of employee representation on its board. In exchange for wage and benefit concessions, Chrysler had granted the UAW president a seat on its board in 1979. Three weeks after the merger announcement, the leaders of IG Metall and the UAW proposed that a UAW representative have one seat on DaimlerChrysler's supervisory board and the company accepted. The two unions also created a "world employee committee" and persuaded the DaimlerChrysler board to engage with it (EIRO, DE9805264N, 7 May 1998; Piazza 2002: 127). These two steps increased the contact between the UAW and IG Metall and DaimlerChrysler works council leaders. The two unions also agreed to promote the organization of the Mercedes plant in Vance, Mississippi.

The DaimlerChrysler merger never produced the promised synergies and Chrysler's performance steadily deteriorated. In 2007, Daimler paid the hedge fund Cerberus Capital Management $650 million to take Chrysler off its hands (*Time*, 30 April 2007). The sale of Chrysler returned the engagement of German automobile producers in the United States to the *status quo ante* with one caveat: BMW and Daimler were producing far more cars in the United States than they did a decade earlier. When it became known in December 2009 that Daimler planned to produce one of its top-selling vehicles, the C-class sedan, in Mississippi, several thousand employees at Daimler's Sindelfingen plant walked out in protest (AL.com, 1 December 2009). The threat suddenly became real to German employees, union officials and works councilors that Daimler could ship to Germany cars from its US plant that were identical to those made in Germany and thereby undermine employment and wages in the German automobile industry.

General Motors bankruptcy in the global financial crisis of 2008-2009 increased cooperation between IG Metall and the United Auto Workers because the UAW's secretary-treasurer, Bob King, held an employee seat on the supervisory board of GM's German subsidiary, Opel. When Bob King became UAW president in 2010, he made union revitalization his top priority (King 2010). UAW membership had fallen below 400,000, which was one-quarter of its 1979 peak. King asserted that for the UAW to survive, it needed to organize foreign-owned automobile factories. One of King's first moves was to seek the advice and support of IG Metall leaders to help the UAW organize the German automobile producers operating in the United States. "Indeed, King and the UAW regarded the institutional power of the IG Metall as a 'game changer' in this struggle" (Fichter 2017: 2).

The reaction to King's overture within IG Metall varied. Some saw no particular urgency or "organizational logic" to "rescue" the UAW, and judged the UAW as too weak and inept to be a viable partner (Fichter 2017: 2). Yet, as momentum grew for talks to create a EU-USA free-trade zone and Volkswagen executed plans to produce again in the United States (this time, at a plant in Chattanooga, Tennessee), concerns grew among IG Metall officials that the United States was becoming a low-cost production site that German companies could use to extract concessions at home (*Nouvelle Europe*, 5 November 2012). IG Metall became more willing to cooperate with its American counterpart.

In 2011, IG Metall engaged in unprecedented ways to help the UAW with its new organizing drives at the Chattanooga and Vance plants. Exchanges between American and German Mercedes employees began in August 2011 when a group of German works councilors and staff met in Alabama with employees from the Vance plant. In February 2012, Kirk Garner, a Vance employee, attended the annual meeting of the Daimler World Employee Committee as an observer. In the spring of 2012, the UAW and German works council staff held workshops for Vance employees interested in learning more about works councils and German industrial relations. In August 2012, a group of thirteen Mercedes employees who were all IG Metall members visited Vance (*Tuscaloosa News*, 22 February 2014).

The Daimler works council in Germany produced a slick magazine for Vance employees written in English titled, *Spark*. A graphic appeared throughout the publication. It was an abstract depiction of a bird with the colors of the US flag on its right wing and those of the German flag on its left wing. The bird hovers over the logos of the UAW, IG Metall and Daimler works council. The graphic included the slogan in German, "*Gemeinsam für ein gutes Leben*," which IG Metall had begun using a few years earlier, and just below it an English translation: "Together for a better life." IG Metall's newspaper for the Sindelfingen plant had a story about the trip (*Brennpunkt*, November 2012). In late January 2013, twelve Vance employees who were members of the organizing leadership council went to Sindelfingen for a week to meet with Daimler and IG Metall representatives, and to see how German industrial relations worked in practice at Daimler (See: https://youtu.be/hjZxhwPgVKQ, accessed 22 November 2016). A second issue of *Spark* – again produced by the Daimler works council in Germany – came out in February 2013. In it, Bob King expressed his gratitude for, "the unprecedented support and involvement of the German union IG Metall and the Daimler Works Council." As the unionization drive continued over the summer of 2013, some Germans returned to Alabama to help (www.AL.com, 23 August 2013). Intervention by German employee representatives, however, changed neither the course of events nor the outcome. UAW organizers were never been able to persuade more than a third of the Vance employees to sign union authorization cards. So, the organizing drive stalled.

Cross-Atlantic exchanges and video conferences also began in 2011 with union sympathizers from the newly opened Volkswagen Chattanooga plant. Horst Mund, head of IG Metall's international affairs department, made frequent visits to both Chattanooga and Vance. Chattanooga employee Justin King participated as an observer at the Volkswagen Global Works Council in November 2011. In June 2012, then

General Secretary of the VW enterprise works council Frank Patta travelled to Chattanooga and met with Volkswagen employees (*Automotive News*, 22 February 2014).

In March 2013, the UAW stepped up its organizing drive in Chattanooga, which included a prominent role for IG Metall. The UAW and IG Metall produced a joint brochure titled, *Co-determining the Future. A New Labor Model.* The professionally produced twenty-page booklet had the logos of the UAW and IG Metall on the cover. It included the same article on "The 21st century UAW" that was used in the first issue of *Spark* from the Daimler organizing drive along with a brief introduction to IG Metall, a letter from then IG Metall Chair Berthold Huber urging Chattanooga employees to join the UAW, and an article from IG Metall international affairs department head Horst Mund praising codetermination. IG Metall also sent Huber's message separately as a stand-alone letter to all 2,350 Volkswagen employees. As the union recognition election at Volkswagen drew nearer, Berthold Huber – who by then had stepped down as IG Metall Chair but remained President of IndustriALL – sent a letter dated December 18 to the employees at VW's Chattanooga plant advocating once again for the UAW and a works council.

A central plank of the UAW's organizing drive at Volkswagen was its willingness to accept a works council, which the union had never done before. On 27 January 2014, the Volkswagen Group of America and the UAW released a joint document spelling out the particulars regarding a works council plus a commitment to "maintaining and where possible enhancing the cost advantages and other competitive advantages that VWGOA enjoys relative to its competitors in the United States and North America, including but not limited to legacy automobile manufacturers" (Volkswagen Group of America and United Auto Workers 2014: 11). On 14 February 2014, the UAW lost the recognition election at Volkswagen Chattanooga, 626 to 712. Heavy intervention by the Tennessee Republican establishment and Washington anti-union groups, the UAW's failure to develop sufficient local community support, and the union's pre-election commitment to maintain and enhance competitiveness – which many workers interpreted as wage restraint – all contributed to the union's loss (Silvia 2018).

The cooperative efforts to organize German transplants in the United States were innovative. IG Metall as well as the Daimler and VW works councils became far more deeply involved in organizing drives in the United States than they ever had before. The efforts have consumed considerable time and resources on all sides. They have altered the course of events, at least in the case of Volkswagen Chattanooga, but they so far have been fruitless. The larger political framework of anti-unionism in the southern United States made it extraordinarily difficult for the exported power of IG Metall and the works councils to change the outcome at the large German automobile assembly plants in the United States that have received sizable subsidies from state and local governments.

Despite the setbacks, IG Metall leaders have not given up on helping the UAW organize in the United States. They have instead shifted focus. In 2015, IG Metall's executive committee approved the "transnational partnership initiative" (TPI), which was a multi-year commitment to support organizing in the United States and Hungary. Then IG Metall Chair Detlef Wetzel justified the undertaking with the observation, "We have to think about value chains in an international context. We therefore will strengthen the internationalization of our union work." To secure jobs and labor standards, IG Metall must "counter cutthroat competition together with our sister unions abroad" (Deutscher Gewerkschaftsbund, Press Release, 26 January 2015; Hübner 2015).

On 19 November 2015, IG Metall and the UAW founded the Transnational Labor Institute (TLI) in Spring Hill, Tennessee. The mission of the TLI is to promote cooperation between IG Metall and the UAW to "improve wages and working conditions for employees at German-owned auto manufacturers and suppliers in the U.S. South," and to "expand on the principle of 'co-determination' between management and employees by establishing German-style works councils or similar bodies to promote employee representation" (UAW, 7 December 2015). The UAW earmarked $2.9 million for the three-year project, which IG Metall leadership supplemented with €1.5 million (IG Metall Vorstand 2015).

The TLI has one full-time employee – Carsten Hübner, a German national – working with the UAW in Spring Hill, Tennessee, to help unionize German parts suppliers. The UAW has scored some successes. In 2016 and 2017, the union won representation elections at a Kirschhoff and an Eberspächer parts plant in Michigan with 100 and 370 employees respectively, despite management opposition and the use of anti-union law firms (*direkt*, November 2011, 5; and IG Metall, Press Release, 12 December 2017). IG Metall executive committee member Wolfgang Lemb observed that, "The Kirschhoff example shows what transnational union work can accomplish in an environment hostile to unions" (IG Metall, Press Release, 5 October 2017). Perhaps, but given that there are over 20,000 employees at German automobile supplier subsidiaries in the United States, the pace of organizing would have to accelerate substantially if the intervention of German employee represenatives is to have a perceptible impact on the sector.

IG Metall Efforts to Assist the Hungarian VASAS Mechanical Engineering Union

Hungary is a good comparison case to the United States to assess the capacity of German employee organizations to export power because many of the same issues arise, but they are in an economic and political system that is starkly different from those of the United States. That said, both cases are examples of how weaknesses in an employment relations system and inhospitable political regimes can severely restrict the effectiveness of attempts by German employee representatives to export power.

Audi, which is a subsidiary of Volkswagen, was the first German automobile firm to invest in Hungary. It opened a motor assembly plant in Győr in 1993 that quickly became an important supplier for Audi throughout Europe. German auto-parts producers soon followed. The next big step came in 2012 when Daimler opened a plant in Kecskemét to assemble C-class models. This plant's potential direct threat to employment in Germany was obvious to employees and their representatives (*Brennpunkt*, December 2011).

German employee representatives have been working with their Hungarian counterparts since the end of the Cold War, but Hungary's employment relations landscape complicates matters. Analysts call Hungary the "land of a thousand unions" (Girndt 2013: 2). Fragmentation produced a motley labor movement riven with rivalries. Adding to the difficulties has been a steep decline in union membership, particularly in the private sector (Makó 2010: 5). Collective bargaining has also been a challenge because there are few employers associations. As a result, collective agreements are mostly with single firms. Hungary enacted a work council law in 1992, but the councils are not nearly as powerful as their German counterparts (Girndt 2013: 5). The larger political environment has also become more difficult in Hungary since the return of Viktor Orbán as prime minister in 2010. Orbán has pursued a right-wing populist agenda that has eroded democracy and undercut labor laws and regulations, which in turn has allowed Hungary to become "a test laboratory" for practices like 12-hour shifts that boost profitability but degrade labor conditions (*Mitbestimmung*, June 2017).

IG Metall has worked with the VASAS mechanical engineering union throughout the post-Communist era. IG Metall leaders deepened the union's cooperation with VASAS by committing €2.3 million for joint projects over three years as a part of the transnational partnership initiative (IG Metall Vorstand 2015). The two unions established joint offices in Győr in 2016 and Kecskemét in 2017. Their purpose is to teach local trade unionists "the ABCs of organizing" and union administration (*Mitbestimmung*, June 2017). In the first seven months of operation, VASAS has organized one additional workplace in the region around Győr and membership has risen by 500 since the joint office opened there (Niemann-Findeisen and Varga 2017). In other words, IG Metall's cooperation with VASAS has had at best a minor impact. It will require a substantial acceleration of organizing in this region with close to half a million citizens to have an appreciable effect on wages and working conditions.

Siemens

Siemens provides some useful variation as a case because it is outside of the automobile sector. Siemens is a large diverse German mechanical engineering conglomerate. For several decades, the company has expanded its presence in the United States largely by buying existing businesses. As a result, Siemens USA has an eclectic portfolio of investments with over 50,000 employees and unions have organized approximately five percent of them. For many years, Siemens had no systematic employee representation in the United States. Lothar Adler, head of Siemens general works council from 2008 to 2014, relied on Wayne Cupp, president of Local 84765 of the International Union of Electrical Workers-Communications Workers of America (IUE-

CWA) to represent all Siemens USA employees, even though Cupp's local only had about 100 members at a Siemens plant in Norwood, Ohio.

Siemens management has maintained a cooperative relationship with IG Metall. In July 2012, Siemens management signed a Global Framework Agreement with the company's general works council, IG Metall and IndustriALL Global Union. The agreement includes guarantees of employees' right to "freedom of association and the right to collective bargaining." At the same time, however, Siemens local management at a small facility in North East, Maryland, hired a law firm that specializes in union avoidance to advise them. On 7 September 2012, the United Steelworkers of America lost the recognition election there, 15 to 24. This loss prompted union representatives from Siemens facilities in the United States to meet later that year with the Siemens enterprise works council and IG Metall representatives in Houston. In the following year, a delegation of Siemens American employees visited Germany.

In 2014, the International Brotherhood of Electrical Workers (IBEW) held an organizing drive at a small Siemens Medical Services facility based in Beaverton, Oregon. Once again, local Siemens managers hired a law firm that specializes in union avoidance. This time, however, the local IBEW officials contacted IG Metall officials and the German union was able to persuade Siemens management to drop the union-avoidance firm and to accept the bargaining unit proposed by the union. The IBEW won the representation election, 13 to 6 (nwLaborPress.org, 2 September 2014).

The Maryland and Oregon incidents led to a meeting of representatives of Siemens general works council, IG Metall and management in 2015 at which the parties agreed to draft a codicil to the company's global framework agreement to cover the United States. The German employee representatives insisted that US unions be included in the drafting process.

In February 2016, IG Metall invited representatives from ten US unions that had organized at least one Siemens facility to a gathering in Orlando, Florida. Three topics were on the agenda: (1) An overview of Siemens USA and its place in the larger company, (2) employee and union experiences with Siemens USA, and (3) next steps for approaching Siemens management. Five of the ten unions sent representatives: the International Association of Machinists, IBEW, IUE-CWA, UAW and United Steel Workers of America. Lee Vickers, head of employee relations for Siemens USA was also there. Vickers indicated that he and his staff would produce of draft letter of understanding for a later meeting. The five unions formed a steering committee after the meeting (Siemens Dialog der IG Metall, 13 July 2016).

On June 13 and 14, a larger gathering with representatives from all of the parties met in Washington, D.C. The first day's meeting was contentious. Günther Erb, head of human resources at Siemens, presented a draft of the letter of understanding. In response, the IBEW representative struck a conciliatory tone, but representatives from the other unions were sharply critical of the draft. They complained that the draft did not include the US trade unions as signatories; it only included the signatories of the Global Framework Agreement, which were Siemens management, the Siemens general works council and IG Metall. Second, the draft required the AFL-CIO and IndustriALL to sign it as a *fait accompli*. Third, the draft did contain provisions for Siemens to take a neutral position regarding unionization during an organizing drive at a Siemens facility, but settling on the text initially proved difficult because the five union representatives at the table differed when it came to the specifics of a neutrality agreement. The second day was more productive. Siemens management representatives agreed to work with the union steering committee and expressed a willingness to compromise. The parties ultimately produced a neutrality agreement (Siemens Dialog der IG Metall, 13 July 2016).

The Siemens case is noteworthy because German actors significantly changed the course of events by taking a strategic approach. Without the German actors, there would have been no letter of understanding. The German actors took the lead to bring together the five US unions and saw through to the end the drafting of an acceptable letter of understanding. It is worth noting that the expenses involved in the Siemens undertaking are far lower than the multimillion dollar commitment that IG Metall's Transnational Partnership Initiative.

To date it is difficult to judge the efficacy of the Siemens letter of understanding. Since the parties signed the neutrality agreement, only a few unionization drives have begun at unorganized Siemens USA facilities and they are only at preliminary stages. Still, Siemens USA is the one case for which intervention by German employee representatives altered events and may even eventually lead to a change in outcome when it comes to unionization at the company.

T-Mobile USA

The case of the Communications Workers of America's long-running organizing campaign at T-Mobile USA, Deutsche Telekom's U.S. mobile-phone subsidiary, adds to the analysis because it involves a different German union – ver.di – and a company that principally sells services rather than goods. The motivations behind ver.di's engagement with a US union differ from those of IG Metall. Deutsche Telekom employees and ver.di officials do not have to worry about T-Mobile USA shipping products to Germany that would undermine employment or compensation. They do worry about the importation of anti-union business practices from T-Mobile USA that run counter to the postwar German tradition of cooperative employment relations known as "social partnership" making their work more difficult. Ver.di officials have balanced a desire to practice transnational labor solidarity and to call out Deutsche Telekom's hypocrisy of embracing social partnership at home while flouting it abroad, with the imperative of focusing on the concerns of dues paying members and preserving the financial well-being of a company that provides jobs and wages.

The Communications Workers of America supported Deutsche Telekom's 2001 entry into the US market through the acquisition of VoiceStream Wireless because DT had a long track record of working cooperatively with unions in Germany. The CWA quickly began organizing efforts at the company, which DT renamed T-Mobile USA. Rather than bring German cooperative labor practices to the United States, Deutsche Telekom management chose to continue VoiceStream's anti-union practices (Logan 2009: 5).

It is common for large German corporations to negotiate a global framework agreement with the national and global union in their sector. Deutsche Telekom instead unilaterally adopted a social charter in 2003, which is still in force, that commits the company to "recognize the right to collective bargaining with the scope of national regulations and existing agreements" (Deutsche Telekom 2017). In the same year, however, T-Mobile USA commissioned a prominent "union avoidance" law firm to write a manual for managers on how to undermine union organizing drives. In 2008, T-Mobile USA replaced the manual with a confidential human resources memorandum, but the new document had the same purpose (Logan 2009: 11-18).

In 2009, the CWA and ver.di established a joint organization called TU-Union (TU). TU is "a non-geographically fixed 'local'" for T-Mobile employees. TU's purpose is:

- to intensify communications and cooperation between CWA and ver.di.
- to set up a platform to be used by the employees of T-Mobile in the United States and Germany to share their experiences.
- to implement a neutrality and anti-discrimination agreement with regards to employee efforts to organize in a trade union at T-Mobile USA.
- to formally recognize trade union representation at T-Mobile USA.
- to support American union members who work for T-Mobile in Germany and German union members who work for T-Mobile in the United States. (ver.di and CWA 2009)

TU has served as a body to bring together pro-union T-Mobile USA employees and to build connections among grass-roots ver.di members as a part of the ongoing unionization campaign.

Unlike IG Metall, which has been strategic, proactive and willing to invest significant resources to assist the UAW organize German automobile producers and parts suppliers in the United States, ver.di has been reactive and unwilling to spend much on the T-Mobile USA campaign. Ver.di is ill equipped for international work. It has no international affairs department and currently allocates only €10,000 from its annual budget to support international undertakings. In something of a mirror image of IG Metall's Transnational Partnership Initiative, the CWA pays the salary of an employee who works at ver.di headquarters in Berlin to pursue matters involving T-Mobile USA (Interview, Ado Wilhelm, 4 July 2017).

To be sure, a small number of dedicated individuals in ver.di have taken the lead to work with the CWA. ver.di's telecommunications newsletter, *Komm*, regularly runs stories about T-Mobile USA and ver.di officials helped to instigate a complaint against T-Mobile USA's anti-union actions at the Organization for Economic Cooperation and Development (International Trade Union Confederation, Press Release, 12 July 2011). There has also been a minor inquiry and a petition in the Bundestag (Deutscher Bundestag 2013 and 2015) as well as a letter to the Chancellor from the heads of the German and US labor movements (Letter, Reiner Hoffmann and Richard Trumka to Angela Merkel, 17 May 2017). Like IG Metall and the UAW, there have also been exchanges of local activists and meetings between the heads of ver.di and the CWA. The International Trade Union Confederation in Brussels maintains a website called "We Expect Better" that is critical of DT and T-Mobile, and the CWA has produced a number of short videos, including some with prominent ver.di officers and Deutsche Telekom works council members expressing to T-Mobile USA employees solidarity and support (e.g., https://www.youtube.com/watch?v=v_mvt97ByLY). Still, these undertakings have been piecemeal. They more closely resemble defensive guerilla warfare than a strategic effort.

Despite repeated urging from CWA leaders, employee representatives at Deutsche Telekom have not used their positions that are the product of codetermination laws to press DT management aggressively to change T-Mobile USA's anti-union activities. The Deutsche Telekom works council leadership understands its mandate to extend only to DT's employees in Germany. Employee representatives on DT's supervisory board have hesitated to force the issue. One explanation for this reticence is the success of T-Mobile USA's chief operating officer, John Legere. Legere is an iconoclast. He has long hair and his signature outfit is a leather jacket over a pink T-Mobile T-shirt. He has been an equally iconoclastic manager. He shattered several industry practices – for example, the use of two-year contracts – thereby transforming T-Mobile USA into the most profitable DT unit. DT CEO Tim Höttges has gone as far as to declare that "the unbelievable success of T-Mobile is the greatest turnaround in the history of the telecom industry" (*Wirtschaftswoche*, 30 June 2017).

Legere's willingness to break the rules extends to US labor law. In 2015, T-Mobile set up an organization called T-voice as an alternative to CWA and ver.di's TU. Two years later, a National Labor Relations Board administrative law judge ordered T-Mobile USA to dissolve T-voice, declaring it to be an illegal company union. Legere responded defiantly: "This is ludicrous. We're appealing. Listening to front-line employees tell us what customers need is imperative to our business. Being the #Uncarrier is all about listening to customers and solving pain points. That's what T-voice does & we will always defend them!" (*FierceWireless*, 4 April 2017). Neither Deutsche Telekom management nor the supervisory board have intervened.

ver.di's makeshift efforts, which have little financial undergirding and have not included full use of the options available through the institutions of codetermination, have thus far not appreciably affected either the course or the result of CWA's effort to unionize T-Mobile USA. After almost two decades, the CWA has organized just two workplaces that employ 30 of T-Mobile USA's 50,000 employees (*Süddeutsche Zeitung*, 20 July 2017).

Conclusion

European economic integration and globalization has facilitated an outward expansion of German foreign direct investment that has left German employees and their representatives increasingly vulnerable to German managers using sites with lower production costs – particularly in parts of the world where employee organizations are significantly weaker – to undercut the bargaining position of German employees in their domestic market.

German trade unionists concluded that in some countries where German employers have built factories, the foreign trade unions are too weak to engage the German firms as an effective interlocutor. They fear the result will be a loss of influence at home because German firms will be able to play off these different sites against each other. Consequently, German trade unionists have attempted in recent years to shore up the power resources of weaker trade union movements by exporting some of their own. German trade union leaders in the mechanical engineering sector have endeavored to use a common transparent objective (i.e., a target for wage settlements in collective agreements that equals the inflation rate plus productivity growth) for the regions geographically surrounding the Federal Republic in order to remove compensation growth as a factor of competition among firms. There are few tangible results of this effort, however. The subsequent euro crisis reduced the undertaking to transnational information exchanges among collective bargaining experts that affect neither the course of events nor the outcomes at the bargaining table.

German trade union leaders have also experimented with exporting power to countries with weaker trade union movements, such as the United States and Hungary. The hope is to bolster foreign trade unions with additional resources and influence in order to expand the coverage of collective bargaining, increase compensation and thereby ward off cheaper imports undercutting German collective agreements. The effort has not been solely material. German employee representatives have also tried to export ideas like social partnership and codetermination. The cases of the US and Hungarian automobile sectors show just how difficult it can be for employee representatives to export power, ideas and institutions to less hospitable environments. Dominant political actors and trends abroad can easily swamp such undertakings. The case of Siemens suggests, however, that in instances of extreme fragmentation, an outside actor with a more strategic vision and some resources can make a difference by bringing together local actors who otherwise would remain isolated from each other.

T-Mobile USA is an example of perhaps the most common type of case found in transnational employment relations. Ver.di's efforts to assist the CWA are neither strategic nor well-funded. They are the product of a handful of individuals pricked by their consciences and prodded by the CWA's calls for solidarity. German employee participants in the highest institutions of codetermination at Deutsche Telekom have hesitated to use their power to change the behavior of T-Mobile management because they see US employees as outside of their jurisdiction and they are reserving their political capital for other purposes. They also fear being blamed for killing the goose that is laying the golden eggs.

All this suggests that, despite recent advances in communication, employee organizations like trade unions and works councils have not managed to overcome the disadvantages they have within capitalism *vis-à-vis* firms when it comes to organizing for collective action (Offe and Wiesenthal 1980). Even the most powerful employee organizations remain very much national organizations within an increasingly globalized world. As such, they have largely been unable to address effectively imbalances that are threatening to undermine their power.

Endnote

[1]This examination focuses only on trade unions in the Federal Republic of Germany before German unification because the structure and activities of the unions of the German Democratic Republic have no explanatory impact on the international activities of the trade union movement in unified Germany for the topic under investigation here.

References

Deutscher Bundestag. 2013. Kleine Anfrage. 17. Wahlperiode. Vereinigungsfreiheit auch bei Tochterunternehmen deutscher Unternehmen sicherstellen. Drucksache 17/12588. Berlin, February 27.

Deutscher Bundestag. 2015. Petition 59803. Deutsche Telekom AG - Massnahmen der Bundesrepublik Deutschland als Anteilseigner im Hinblick auf die Einhaltung von Arbeitnehmerrechten beiUnternehmen im Ausland, Berlin, 6 July.

Deutsche Telekom. 2017. *Sozialcharta.* Bonn: Deutsche Telekom.

Dufresne, Anne. 2002. "Wage Co-ordination in Europe: Roots and Routes." In Philippe Pochet, ed., *Wage Policy in the Eurozone.* Brussels: Lang, pp. 79-109.

Ehrenberg, Herbert. 2008. "Aktive Lohnpolitik. Viktor Agartz und die gewerkschaftliche Lohnpolitik in den 1950er und 1960er Jahren." In *Wirtschaftsdemokratie und expansive Lohnpolitik. Zur Aktualität von Viktor Agartz*, edited by Reinhard Bispinck, Thorsten Schulten and Peeter Raane. Hamburg: VSA, pp. 66-81.

Fichter, Michael. 2004. "Normalizing German-American Labor Relationships in a Changing International Environment." In *The United States and Germany in the Era of the Cold War, 1945–1990. A Handbook, vol. 2: 1968–1990*, edited by Detlef Junker. Cambridge: Cambridge University Press, 2004, pp. 460-65.

Fichter, Michael. 2017. *Transnationalizing Unions: The Case of the UAW and the IG Metall.* Berlin: Friedrich Ebert Stiftung.

Girndt, Rainer. 2013. "Ungarns Gewerkschaftslandschaft in Bewegung. Der Versuch einer Konzentration der Kräfte." Berlin: Friedrich Ebert Foundation.

Gollbach, Jochen, and Thorsten Schulten. 2000. "Cross-Border Collective Bargaining Networks in Europe." *European Journal of Industrial Relations*, Vol. 6, no. 2, pp. 161-179.

Höpner, Martin, and Martin Seeliger. 2017. "Transnational Wage Coordination as a Means of Stabilizing the Euro? Has Not Worked in the Past, Does Not Work Now, Will Not Work in the Future." Presented at Rethinking German Political Economy: Lessons for Comparative Theorizing after the Social Democratic Century, Goethe Institute, San Francisco, CA, 30 August.

Hübner, Carsten. 2015. *Globale Wertschöpfungsketten organisieren. Eine Herausforderung für Gewerkschaften.* Berlin: Friedrich Ebert Foundation.

King, Bob. 2010. "A UAW of the 21st Century." Center for Automotive Research Conference, University of Michigan, Ann Arbor, MI, August 2.

Logan, John. 2009. *Lowering the Bar or Setting the Standard? Deutsche Telekom's U.S. Labor Practices.* Washington, D.C.: American Rights at Work Education Fund.

Makó, Csaba. 2010. *Gewerkschaften im erstarkenden Kapitalismus – Das Beispiel Ungarn.* Budapest: Friedrich Ebert Foundation.

Niemann-Findeisen, Sören, and Marika Varga. 2017. "Automobil- und Zuliefererkonferenz der IG Metall, Bezirk Mitte. Niedernhausen, 4 May.

Offe, Claus, and Helmut Wiesenthal. 1980. "Two Logics of Collective Action: Theoretical Notes on Social Class and Organizational Form." *Political Power and Social Theory*, No. 1, pp. 67-115.

Piazza, James A. 2002. *Going Global: Unions and Globalization in the United States, Sweden, and Germany.* Lanham, MD: Lexington Books.

Pond, Richard. 2009. "A Regional Perspective on Coordinating Collective Bargaining Policy in the Energy Sector." *Transfer*, Vol. 15, no. 2, pp. 315-19.

Pulignano, Valeria. 2010. "Syndicats et coordination des négociations collectives en Europe." In *Relations collectives de travail en Belgique*, edited by M. Capron and E. Leonard. Brussels: CRISP, pp. 437-56,

Schulten, Thorsten. 1998. "Tarifpolitik unter den Bedingungen der europäischen Währungsunion. Überlegungen zum Aufbau eines tarifpolitischen Mehr-Ebenen-Systems am Beispiel der westeuropäischen Metallindustrie." *WSI Mitteilungen*, Vol. 51, no. 7, pp. 482-93.

Schulten, Thorsten. 2002. "Europeanisation of Collective Bargaining: An Overview on Trade Union Initiatives for a Transnational Coordination of Collective Bargaining Policy." WSI Discussion Paper No. 101, Düsseldorf, May, pp. 2-30.

Silvia, Stephen J. (2013) *Holding the Shop Together: German Industrial Relations in the Postwar Era.* Ithaca, N.Y.: Cornell University Press.

Silvia, Stephen J. (2016) *Organizing German Automobile Plants in the USA: An Assessment of the United Auto Workers' Efforts to Organize German-Owned Automobile Plants.* Düsseldorf: Hans-Böckler-Stiftung.

Silvia, Stephen J. (2018) "The United Auto Workers Attempts to unionize Volkswagen, Chattanooga." *Industrial and Labor Relations Review.* Vol. 71, no. 3 (May), pp. 600-24.

Slobodian, Quinn. 2010. "West German Labor Internationalism and the Cold War." In *Divided But Not Disconnected: German Experiences of the Cold War*, edited by Tobias Hochscherf, Christoph Haucht, and Andrew Plowman. New York: Berghahn, pp. 77-89.

ver.di and Communications Workers of America. 2009. "Agreement between ver.di and CWA to Establish and Jointly Affiliate TU-Union." Washington, D.C.

Vis, Barbara, and Jaap Woldendorp. 2015. "Same as It Ever Was? The (Lack of) Influence of European Integration on Corporatism in the Low Countries." In *European Integration and Consensus Politics in the Low Countries*, edited by Hans Vollaard, Jan Beyers and Patrick Dumont. London: Routledge, pp. 155-173.

Volkswagen Group of America and the United Auto Workers. 2014. "Agreement for a Representation Election." Unpublished document. Chattanooga, Tennessee, January 27.

V. 2018 LERA Lifetime Achievement Award

Remarks

MARCIA L. GREENBAUM
Mediator/Arbitrator

Thank you for that generous introduction. I recall the story told by my dear mentor, Cornell Professor Jean T. McKelvey, whom I greatly admired, and to whom I pay tribute. She was called upon to give a lecture at a luncheon and she went on and on. Finally, she noticed that the audience was getting restless. People were moving in their seats, coughing and starting to whisper. She apologized for talking so long, explaining that she had forgotten her watch and that there was no clock on the wall. At that time, an audience member piped up: "But Professor McKelvey, there's a calendar on the wall!" I note there isn't a calendar on the wall in this room.

I'm very grateful for the recognition LERA has bestowed upon me with this Lifetime Achievement Award. I am also honored to be in the good company of those past and present who have received this award. I extend special thanks to Harry Katz (given his many hats, I'm not sure whether to address him as Professor Katz, Dean Katz, Provost Katz or President Katz), who was instrumental in nominating me. I also want to express my gratitude to the Cornell ILR School, where I was both an undergraduate and graduate student and gained the knowledge to pursue my career.

On the other hand, I have mixed emotions about receiving this award. It is the use of the word "lifetime" that gives me pause.

I feel very lucky to have been able to have a 50+ year career as an arbitrator, where decisions make some winners and others non-winners—a challenge if one wants to maintain acceptability. I see an arbitrator much like an umpire in a baseball game. It does not matter who the pitcher is or who the batter is—it is either a ball or a strike. I also serve as a mediator in a process where I can bring disputing parties together and help them reach a settlement.

Both forms of conflict resolution ensure that there is some fairness and justice in the workplace by having a neutral weigh in on the matter in dispute. It is the right of employees, through their union, to have a voice, to be heard, and to get the benefits of their labor in a safe environment. On the other hand, the employer has a right to have an employee who will give a fair day's work as all in the workplace have an interest in the employer staying in operation. Thus, both parties also have obligations.

The job of the grievance arbitrator is to hear both parties, and issue a decision that takes into account the employee and the union's claims, management's position, the facts, the relevant contract language, the law of the shop and sometimes the law of the land. These processes, as substitutes for strikes and courts, contribute to holding the social fabric together. Hopefully, the U.S. Supreme Court will recognize in the *Janus* case that strong unions play an important role in a democratic society, and allow them to continue to collect fair share dues fees from those who are not members, but who benefit from the fruit of the union's bargain.

Some years ago, I served on a tripartite board to recommend minimum wage rates for the lingerie industry in Puerto Rico that was exempt from the higher rate set by the 1938 Fair Labor Standards Act, which is having an 80th anniversary this year. As it was easy for the employer to pick up his sewing machines and relocate to countries where labor is cheaper, it was important that the rate not be set so high as to encourage the employer to move off the island, and cause these, mostly women, workers to lose their jobs. After going back and forth talking with the union and management representatives, I concluded that the union position was too high and I voted with management.

A representative of a well-known undergarments manufacturer, who was serving on that tripartite board, offered to have me visit their showroom in New York City, where he would provide me with luxury lingerie. I said, "No, thank you. I don't do that." He repeated his invitation to "outfit" me. I said, "That is not necessary. I voted my conscience." He replied, "You don't understand. I am offering to make you the best 'undressed' woman in Boston." I replied, "You don't know that I am not that already."

Thank you for this much appreciated award.

Remarks

MORRIS M. KLEINER
University of Minnesota, Federal Reserve Bank of Minneapolis, Upjohn Institute of Employment Research, and the National Bureau of Economic Research.

I am greatly honored to receive the 2018 LERA Lifetime Achievement Award; this award truly came as a surprise to me and I am deeply grateful to follow in the large footsteps of luminaries in our field that included professors, who were also practitioners, such as John Dunlop, Ray Marshall, Clark Kerr and George Shultz among many others, who have also been so honored. I would like to thank my family who encouraged me to continue to contribute to the field far beyond my expiration date.

My journey in the field was encouraged by Paul Gerhardt who suggested that I join the then Industrial Relations Research Association as a new graduate student at the University of Illinois, and who said that this organization was the true measure of being a professional in the field. My interest in industrial relations at the University of Illinois was spurred further by Paul Hartman, who had written empirical case studies of collective bargaining and productivity in Pacific longshore industry and by Hugh Folk, whose work on modeling the workforce was far ahead of its time. Each taught me the importance of asking important questions with good data and rigorous analysis.

At the University of Kansas as an assistant professor I was mentored by the students of former IRRA President George P. Shultz, such as Joe Pichler and Charles Krider, who encouraged and incentivized me to come to the IRRA annual meetings. During a leave at the Brookings Institution in Washington D.C. I saw the importance of labor economics in helping policy makers think about policy issues in new ways. During a subsequent sabbatical at Harvard I was able to work with Richard B. Freeman, who I saw as the most insightful and imaginative labor economist, and he had just published his best selling *What do Unions Do?* We were able to give our favorite papers at LERA annual meetings. He also cemented my interest in studying labor market institutions and how they influence productivity as the most important issue for economists. A critical aspect of my research was my affiliation with the National Bureau of Economic Research, which further encouraged my interest in the study of the employment relationship.

At the University of Minnesota, I found a home at an institution that valued both public policy and labor studies. Tom Donahue, a long-time LERA member and at the time President of the AFL-CIO, encouraged my analysis of the importance of unions that had evolved from my work with Richard Freeman, Jonathan Leonard, Adam Pilarski, and David Weil. What made the Humphrey School and the Center for Human Resources and Industrial Relations special is the extraordinary colleagues who span a variety of disciplinary backgrounds. Academic research and policy conferences were encouraged from my interactions initially with Mario Bognanno, and then later with John Budd, on the future of the labor movement to labor institutions from an international perspective. The presence of these colleagues has heavily influenced the research and policy work that I have done during my career

I would like to give you a few examples.

My work on empirical case studies has ranged from industrial relations in aviation with Adam Pilarski, shoe manufacturing with Richard Freeman, car parts with Sue Helper, and banking with Ann Bartel and the late Casey Ichniowski. With these case studies some say I could have opened my own strip mall.

The work that has gotten me the most notoriety was on occupational regulation, where I was able to transfer my approach to unions learned at LERA to the study of the costs and benefits of government regulation. The Upjohn Institute and its President Randy Eberts funded and provided a platform for three of my books on the topic. The Federal Reserve Bank of Minneapolis, under the leadership of Art Rolnick and later Mark Wright, supported research, data gathering, policy applications, and conferences on the topic. From studying the influence of licensing on dentists with Bob Kudrle to the aggregate economic effects of occupational regulation with former U.S. Council of Economic Advisers head Alan Krueger, this work has allowed the issue to become one that has had resonance with the top officials in both the Obama and the Trump administration, as well as in the halls of Congress in a nonpartisan manner. The lessons learned at LERA allowed my work to have both a scholarly and public policy influence. LERA stands as a way to meet the challenging new economy and workforce issues ahead. This organization is an excellent way of exchanging experiences and ideas, and it has a unique and vital role to play. I am delighted that LERA continues to grow and prosper for the next generation of researchers and public policy experts as it did for my generation.

VI. Income, Work, and Wealth: UBI, Job Guarantees, Baby Bonds, and Care Work

Modernizing U.S. Labor Standards for 21st Century Families

BRIDGET ANSEL
HEATHER BOUSHEY1
Washington Center for Equitable Growth

ABSTRACT

Women now make up almost half the U.S. workforce. Despite the central role women play in the U.S. economy, our labor laws and institutions do little to address the various ways in which women are held back at work. This not only hampers women's economic well-being, but also has implications for U.S. productivity, labor force participation, and economic growth. In this paper, we propose policies aimed at boosting women's economic outcomes: paid family leave, fair scheduling, and combatting wage discrimination. We show how enacting carefully designed policies in these categories will better address the challenges of today's labor force, enhance women's economic outcomes, and provide benefits for the national economy.

Introduction

The growing number of women in the U.S. workforce over the past 30 years has reshaped both traditional gender roles and the American economy. Since the late 1970s women have outnumbered men in U.S. college enrollment (National Center for Education Statistics 2016), allowing them to break into new occupations that had been dominated by men, and to make more valuable contributions to the economy. Women's paid work boosted U.S. GDP by an estimated 11 percent between 1979 and 2013 (Appelbaum, Boushey, and Schmitt 2014). And as men's earnings fell by 9.5 percent over the past three decades, it was women's increased working hours that kept household income from declining in each income group (Boushey 2016; Boushey and Vaghul 2016; Glynn 2014). With women now making up close to half the nation's workforce, it is clear that their earnings are crucial for families' well-being and the nation's economic strength.

Despite the central role that women play in the U.S. economy today, federal policies and labor laws remain anchored in the past, hampering further progress. Most of the laws that govern labor standards, such as the Fair Labor Standards Act, were enacted as part of the New Deal in the 1930s. While these laws underpin employer and employee relationships, they are predicated on certain outdated premises: jobs are assumed to have predictable, standardized schedules and families are assumed to have a single breadwinner and a stay-at-home caregiver. The latter assumption is explicitly gendered, and is based on a view of the idealized upper- or middle-class white family. In reality, many women in the United States, especially low-income and racial and ethnic minority women, have historically worked to support their families (Frye 2016; Landry 2000).

The limits of federal labor laws can be partly ascribed to the fact that they were enacted at a time when the U.S. labor movement was considerably stronger. Many important labor standards were therefore not directly codified through law, but rather negotiated across the bargaining table. From the New Deal through the early 1970s, unions grew to cover about one-third of the workforce and helped mitigate exploitative labor practices for some workers. But over the past 30 years, private sector labor unions have

Corresponding author's address: Heather Boushey, Washington Center for Equitable Growth, 1156 15th Street NW, Suite 700, Washington, DC 20005 declined, and only about 7 percent of private sector workers are now covered by union contracts (Bureau of Labor Statistics [BLS] 2016a).

Today, the United States is left with labor laws and institutions that do little to address the various ways in which women are held back at work. Policymakers' failure to implement any sort of national paid leave policy forces workers to choose between the loss of wages—or even the loss of their jobs—and taking the time to care for a new child, their own health, or an ill family member. Working hours that are too long, unpredictable, or insufficient can create work-life conflicts that make it difficult to manage a paid job with other responsibilities. On top of work-life conflicts, women are often subject to wage discrimination, and a lack of pay transparency means that this issue often goes unaddressed.

Given the failure to address these issues, it is no wonder that women's economic progress has stalled by several measures, including labor force participation and the gender wage gap. Without policies that address work-life conflicts, many women stop working altogether. Unlike other developed countries, most of which have policies addressing these issues, the United States has seen a decline in women's labor force participation in recent years, especially for women in their 30s and 40s (Goldin and Mitchell 2017).

These labor market outcomes are not inevitable, but are to a large extent the product of deliberate policy choices. Amending U.S. labor laws can enhance women's economic outcomes, thus providing a boost to the national economy through increased productivity, greater labor force participation, and increased demand for goods and services. In this paper we propose design principles for three groups of policies—paid family leave, fair scheduling, and combatting wage discrimination—aimed at boosting women's outcomes.

The Challenge

Women continue to face a host of impediments to their full participation in the labor market.

Many of these impediments have been studied over the course of decades, providing insight into their effects on women's employment and earnings as well as other consequences for families. We classify these challenges into three categories: caregiving responsibilities, burdensome scheduling practices, and wage discrimination.

Caregiving Responsibilities

The United States is one of the only countries in the world without a national paid leave policy. Some U.S. workers are eligible for 12 weeks of unpaid leave through the Family and Medical Leave Act of 1993 upon the birth or adoption of a new child, serious illness, or to care for a family member. But because of the law's eligibility requirements, only 60 percent of workers and about 20 percent of new mothers have access to legally mandated unpaid leave; those who are excluded from access are disproportionately lower income (Council of Economic Advisers 2014) and less educated (U.S. Department of Labor n.d.). Even those who are eligible for unpaid time off, however, often do not take it. A recent survey conducted by the Pew Research Center finds that one in six U.S. workers said they needed to take time off work in the past two years but were unable to do so, primarily for financial reasons (Horowitz et al. 2017).

Access to paid time off is even more limited. In 2016 only 14 percent of the private sector workforce and 4 percent of workers in the bottom tenth of the wage distribution received paid leave through their employer (BLS 2016b, table 32a). A larger fraction has some other access to paid time off: 38 percent of workers in the United States have access to temporary disability insurance to deal with a personal medical condition without losing pay, but most disability insurance does not cover the care of a family member (BLS 2016b, table 16a).

For many families, the birth of a child is associated with a significant decline in financial well-being (Stanczyk 2016). To cope, many families—especially low-income families—go into debt, put off paying their bills, or return to work too early, with negative consequences for mothers and children (Horowitz et al. 2017). Even if parents do not return to work right away, the fall in income around the time of the birth can harm children. Money matters for kids, especially young kids, even when controlling for other family parental characteristics (Sandstrom and Huerta 2013). Furthermore, in other economies, paid maternity leave has a profound effect both on children's long-term development and on their future productivity (Carneiro, Loken, and Salvanes 2015).

The Need for Comprehensive Paid Leave

Discussions of paid leave policy often focus on the needs of new parents, but parental leave is not the only valuable use of paid leave: others include dealing with a personal health problem, caring for a family member with a serious health condition, or addressing needs associated with a family member's military deployment. With an aging population and fewer stay-at-home caregivers, an increasing number of workers need time off to care for a family member or for self-care. In fact, those who take leave are more likely to do so for personal medical reasons or to care for a family member than to care for a newborn child (Horowitz et al. 2017).

Workers who are dealing with a personal or family illness face a unique set of challenges. Caring for an ailing family member often requires intermittent leave, taken in small time increments to, for example, take someone to the doctor or spend an afternoon providing care. While more research is needed to determine how differing lengths of family and medical leaves affect individual and economic outcomes, the limited evidence that does exist shows that giving workers some leave for nonparental factors can positively affect both health and labor market outcomes. For example, Earle, Ayanian, and Heymann (2006) observed nurses who had experienced a heart attack: those with paid leave were much more likely to return to work compared to those without this benefit. A study of paid leave in California found that giving workers some time off increases the likelihood that workers—particularly low-income workers—will stay in the labor force following personal and family health events (Appelbaum and Milkman 2011).

Without a comprehensive paid leave program that covers all family care needs, families struggle to address work-life conflicts, and that struggle in turn generates social and economic costs. Half of workers who need leave, but do not have access to it, postpone or never receive critical medical treatment, which has costs for our health-care system as well (U.S. Department of Labor 2015). With the aging of the baby boomer population, a growing number of workers are providing care for an elderly relative. Of these caregivers, seven in ten have had to make work accommodations such as cutting back on hours (and therefore wages) or dropping out of the workforce altogether (Feinberg and Choula 2012). Every worker at some point in their life will need to take time off work for family or health reasons, making this an issue that affects all workers. But women continue to take on the bulk of caregiving responsibilities for children and ill family members, making paid leave particularly consequential for women's participation and success in the labor force.

Although paid leave is especially valuable for many women, its predominate usage by women can negatively affect pay equity. Policies targeted exclusively to women can lead employers to discriminate against young women—even those without children—if employers expect them to use maternity leave (Thomas 2016). In contrast, countries that have more gender-neutral paid leave programs have made bigger strides toward closing the gender pay gap (World Economic Forum 2013). This is partially because men's use of leave frees up women to engage in paid work. For example, in Iceland fathers are given three months of nontransferable paternity leave following the birth of a child. An evaluation of the policy revealed that "the division of care between parents ... has changed in the intended direction and that is mainly due to the law" (Arnalds, Eydal, and Gíslason 2013, 323). In addition, a positive association was discovered between the length of leave that men take and their involvement in care once the leave period is over (Arnalds, Eydal, and Gíslason 2013).

These outcomes aren't specific to Iceland. Quebec's paid paternity leave program had a "large and persistent impact on gender dynamics within households even years after the leave period ended," with fathers taking on a larger share of domestic work and childcare, allowing more time for women to participate in the labor market (Patnaik 2015).

Burdensome Scheduling Practices

Long but irregular work schedules, as well as just-in-time scheduling practices, are a problem for a growing number of workers as they seek to balance work and life commitments. The Fair Labor Standards Act sought to check overwork, yet the failure to update the overtime salary threshold makes the law increasingly ineffective at curtailing long work hours for the vast majority of U.S. workers. While some workers are logging more time at work than ever before, others struggle to get enough work hours to make ends meet (Jacobs and Gerson 2004).

Both of these trends harm women's economic outcomes in particular and family incomes overall. In jobs that require long hours, workers might not be able to address conflicts between their outside and family commitments.

On the other end of the spectrum, part-time jobs that do not provide enough hours can hurt women's ability to provide for their families, especially because workers are likely paid a lower hourly rate than they would be paid in a comparable full-time job. In addition, the rise in unpredictable and nonstandard work hours makes it more difficult to arrange childcare or pursue the education necessary for upward mobility.

Overwork and the Overtime Threshold

Shifts in the way firms organize work over the past 40 years have generated highly demanding jobs characterized by long hours and little flexibility, at least within certain occupations. This trend has created disproportionate difficulties for women, leading them to scale back career aspirations or drop out of the labor force entirely. Stone and Lovejoy (2004) surveyed women who either quit or cut back on hours at their professional or managerial jobs. One-third of those who quit and nearly two-thirds of those who scaled back to part-time work cited long, inflexible hours as the reason (Stone and Hernandez 2013; Stone and Lovejoy 2004).

While some employers pay workers overtime when they ask them to work long hours, that number has dwindled over the past 40 years as the Fair Labor Standards Act's overtime threshold has declined in inflation-adjusted terms. Today, many salaried workers will not earn overtime pay unless they earn $23,660 or less a year. This threshold, which is below the federal poverty level for a family of four, covers only 8 percent of salaried workers, leaving millions of employees without overtime protections.

Overwork has implications for pay equity, increasing the gender wage gap by about 10 percent (Cha and Weeden 2014). Harvard University economist Claudia Goldin cites long hours and the emphasis on office "facetime" as the "last chapter" in attaining gender equality (Goldin 2014).

Underwork and Unpredictable Schedules

While one segment of the working population is spending more time in paid employment, another is having trouble getting enough work to make ends meet. An estimated 5.2 million workers are currently working part time, as shown in Figure 1, but are available for and would prefer full-time employment. Part-time jobs are most prevalent within the low-wage retail and hospitality industries and are disproportionately filled with Hispanic and African-American women. The schedules and hours for many of these jobs are also unpredictable, compounding the financial and emotional burdens that these workers face. Research shows that part-time workers often do not receive benefits and tend to have lower wages than their full-time counterparts, even if they are doing the same job (Golden 2015). A number of cities are considering policies to address this problem: for example, San Jose, California, recently implemented a law aimed at giving part-timeworkers access to more hours, but it is too soon to fully evaluate the law's effect (GovDocs 2017).

Figure 1
Number of People Employed Part-Time for Economic Reasons, 2000–2017

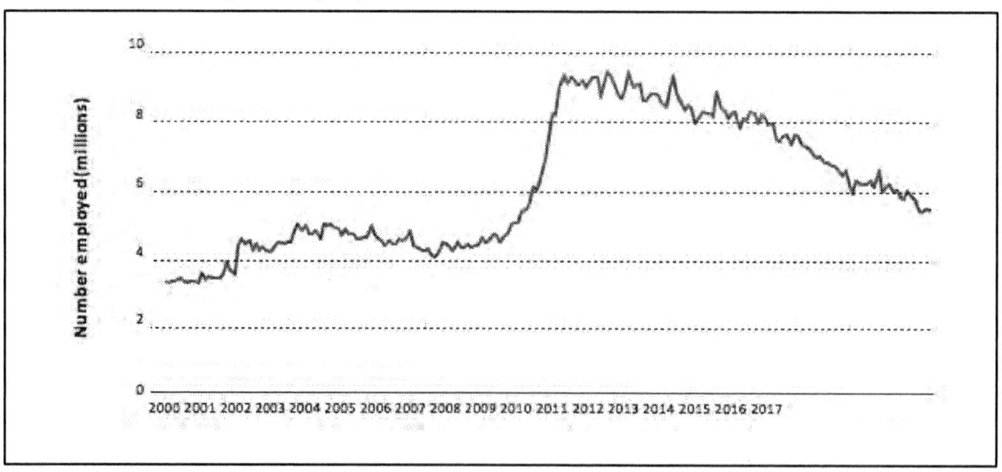

Source: Current Population Survey, 2000–2017.

Note: Montly values are seasonally adjusted.

A growing number of U.S. workers are grappling with unpredictable, constantly shifting schedules. Often aided by "just-in-time" scheduling software, many companies try to generate work schedules based on predictions of consumer demand. But in doing so, they may give their employees only a few days' notice of their schedule for the coming week and require them to remain on call and wait to see if they are needed to work; if sales are slow, employers might send workers home without pay (Boushey and Ansel2016b).

About 17 percent of workers nationally have unpredictable schedules (Golden 2015). While this is a problem for workers of all incomes, jobs with erratic schedules are more heavily concentrated among low-income workers, especially in the retail and service industries (Boushey and Ansel 2016b). See Tsable 1 for information about irregular scheduling by occupation. Research also shows that women, especially women of color, tend to be most affected by these schedules. In fact, more than one-third of female hourly workers in their prime childbearing years receive their schedules a week or less ahead of time (Economic Policy Institute 2017).

Unsurprisingly, schedules that are unpredictable wreak havoc on workers and their families. Earnings fluctuate week to week depending on how many hours employers assign to their employees, making it impossible for workers to predict whether they will earn enough to make ends meet. Unpredictable work schedules are associated with household financial insecurity, even after adjusting for hourly wages and overall income (Schneider and Harknett 2016). Scheduling instability has also been found to be a key driver of the rise of income volatility (Mitchell 2017).

Table 1
Share of U.S. Workers by Shift Type and Occupation, Pooled Years 2002, 2006, and 2010

	Irregular	Split/rotating	Regular
All (n=4,641)	10%	7%	83%
Type of occupation (share of total)			
Executive/admin/management (15%)	9	4	87
Professional specialty (19%)	11	5	84
Technicians (4%)	8	11	81
Sales occupations (11%)	15	10	75
Admin support (clerical) (13%)	4	5	91
Service occupations (16%)	10	14	76
Farming/precision production (11%)	8	3	89
Operators/laborers (11%)	10	9	81

Source: Golden 2015.

Without the time and work predictability required to manage their nonwork commitments, women in particular sometimes have to limit their time engaging in paid work, which results in less income for the family. Some women drop out of the labor force altogether, contributing to a declining female labor force participation rate (Boushey and Ansel 2016a, 2016b). Constantly shifting schedules can also mean that workers are unable to get a second job or go back to school, jeopardizing their longer-term economic stability as well. These kinds of jobs are especially onerous for parents, who can find it difficult to find and keep reliable childcare. It is no wonder that the stress from unpredictable schedules takes a physical toll: Schneider and Harknett also found that those with an unpredictable schedule were more likely to have poorer physical health and to suffer from "serious psychological distress" (Schneider and Harknett 2016, 13).

While some cities have begun passing legislation to limit unpredictable schedules, there is no federal law that prevents employers from requiring employees to work with little advanced notice. That means, for most workers, that the cost of doing business is being pushed onto workers and their families.

Wage Discrimination

Much of the disadvantage suffered by women in the labor market would be addressed by well-designed work-family policies such as paid family leave and fair scheduling. However, wage discrimination still plays an important role in driving gender and racial wage gaps, and would likely continue to do so even after work-family challenges were met. After adjusting for factors like labor force experience, union status, race and ethnicity, and occupation, one recent study finds that 38 percent of the gender wage gap remains unexplained, suggesting that labor market discrimination plays an important role (Blau and Kahn 2017).

This means that even after accounting for observable differences between male and female workers, and white workers versus minority workers, women and people of color still face pay disparities (Schneider and Gould 2016). In fact, almost 60 percent of women would earn more if they were paid the same as men with equivalent levels of education and work hours (Milli et al. 2017). Eliminating pay discrimination through a boost to women's wages would help families and the economy alike: the number of children living in poverty with working mothers would be reduced by 2.5 million and GDP would be higher by $512.6 billion in 2016, a 2.8 percent increase (Milli et al. 2017).

Gender and racial wage discrimination has also been demonstrated in field experiments. Science professors who were given employment applications for a laboratory manager position that were identical except for one part—the candidate's name was shown as either Jennifer or John—made an average starting salary offer to John of $30,000 compared to Jennifer's $26,500 (Moss-Racusin et al. 2012). Similarly, there is the much-cited study where researchers sent out identical resumes and found that applicants with "black sounding" names such as Lakisha or Jamal were less likely to get call-backs than applicants with names such as Emily or Greg (Bertand and Mullainathan 2003). Other research analyzing the job searches of nearly 5,200 unemployed people in New Jersey found that potential employers offered black job applicants significantly less compensation than white job applicants (Fryer, Pager, and Spenkuch 2011).

Even a single instance of this kind of pay inequity can reduce workers' wages throughout their careers. Wage negotiation later in women's careers is unlikely to be of much help, given that women are disproportionately likely to experience negative consequences when they do negotiate, and are consequently unwilling to do so. Men—especially white men—negotiate higher salaries on average without negative effects (Bowles, Babcock, and Lai 2007).

The Persistent Impact of Salary History

Compounding the effects of discrimination is employer usage of salary history. Employers often evaluate candidates and make offers based on what they earned at previous jobs; half of all workers report that their current employer learned at least some of their wage history (Hall and Krueger 2012). To the extent that women and people of color are discriminated against early in their careers, this wage history information can allow past discrimination to follow workers throughout their careers. Because employers sometimes think that salary history is a means to evaluate a worker's productivity, a previous salary that is too low—regardless of whether the low salary level was due to ability or discrimination—could deter employers from making a job offer (Barach and Horton 2017). On the other hand, salary disclosure requirements could also harm older workers who have been laid off or who took time out of the workforce because employers might view them as too expensive.

A number of state and local governments, including those of Massachusetts, New York, and Philadelphia, have responded by banning the employer practice of requiring workers to disclose salary information during the hiring process. While the full effects of these relatively recent bans have yet to be evaluated, the results of a recent field experiment suggest that they might confer benefits (Barach and Horton 2017). Employers who could not see a job applicant's salary history responded by evaluating more applicants overall, asking more questions, and arranging more face-to-face interviews to evaluate an applicant's fit for the job. Employers without access to applicants' salary history also interviewed and hired workers with relatively lower past wages compared to employers who did have access to this information. Not having to disclose their previous salaries also gave applicants more bargaining power; these workers were able to secure higher pay compared to applicants who were required to divulge their pay history (Barach and Horton 2017). Without an applicant's salary history, employers must evaluate and make job offers based on an applicant's tangible skills and experience.

Pay Secrecy and Worker Bargaining Power

When wage discrimination occurs, many women are unaware of the problem due to formal or informal prohibitions of employee discussions about pay. The result is that employers may purposely or inadvertently pay their workers different amounts for the same kind of work. Bans of employee pay discussion appear to be effective at preventing workers from demanding higher pay and wage equality (see box 2). Even if an employee suspects pay inequity, it is difficult to prove pay discrimination without a disclosure or an employment discrimination charge (Rosenfeld and Denice 2015).

Pay secrecy remains common in many U.S. workplaces, and about half of all workers—and more than 60 percent of private sector workers—report that their workplace formally or informally bans workers from discussing their salaries (Institute for Women's Policy Research 2014). Such is the case with Lilly Ledbetter, one of the few female supervisors for Goodyear Tire in Alabama when she worked for the company in the 1980s and 1990s. Ledbetter, who worked at Goodyear for almost two decades, was unaware that she earned less than the other 16 male supervisors because Goodyear prohibited employees from discussing pay. She realized that she was underpaid only after receiving an anonymous note (National Women's Law Center 2013).

Ledbetter's subsequent lawsuit and the action taken by Congress and signed into law by President Obama in January 2009 (The Lilly Ledbetter Fair Pay Act) extended the statute of limitations for filing an equal pay lawsuit. Previously, workers filing a pay discrimination complaint were required to do so within 180 days of the first time they are paid less than their peers. Obviously, this is impossible if employees do not know they are being discriminated against; the statute of limitations now resets with every new paycheck affected by the original discriminatory action. While this is a step in the right direction, the law does not directly address pay secrecy.

The widespread nature of pay secrecy is surprising given that, excluding supervisors and managers, it is illegal under the National Labor Relations Act (NLRA) to prohibit discussions of pay, even informally (Bierman and Gely 2004; Gely and Bierman 2003). However, most employees do not know that pay secrecy is illegal; because there are few penalties for violating the NLRA, employers have little incentive to adhere to the law (Gely and Bierman 2003). Even if the penalties were stronger, the existing exclusion of managers and supervisors is a problem considering that women and people of color are underrepresented in these positions (Warner 2014). Had Ledbetter found out that she was underpaid through discussions with her male colleagues, she would have been violating her company's policy and could have been legally fired (due to her status as a supervisor).

By contrast, employer disclosures of pay information can level the playing field between employers and employees, and help reduce unjustifiable pay gaps. In a recent study of British workplaces, employees who report that their managers are "very good" at disclosing financial information and pay earn 8 percent to 12 percent more than those who report that their managers are "very poor" at sharing this kind of information (Rosenfeld and Denice 2015). In the United States, publishing the salaries of California public employees online compressed the pay of managers by 8 percent, suggesting that pay transparency helped expose and remedy difficult-to-rationalize differences in pay (Mas 2014). Other research shows that within-establishment pay disparities between men and women are smaller in gender-balanced unionized industries, in part because unions often have access to a company's financial information (Elvira and Saporta 2001).

Pay transparency is valuable in part because it motivates employers to create fair pay systems, while also allowing employees to monitor and speak up about discriminatory salary practices. It is also effective in adding legitimacy to workers' salary requests, and makes it difficult for employers to justify differing salaries for men and women doing equivalent work (Rosenfeld and Denice 2015). Salary transparency has an upside for employers as well, reducing worker distrust and boosting productivity. Pay transparency has been shown to help employees collaborate more productively—most likely because workers could more accurately judge their colleagues' skill level by how much they were getting paid (Belogolovsky et al 2016). In addition, workers who know their colleagues' salaries could be motivated to work harder, boosting their output by 10 percent (Huet-Vaughn 2015). By contrast, pay secrecy can actually decrease performance because it takes "a toll on the ability of the firm to retain its best performers" (Belogolovsky and Bamberger 2014, 37).

A New Approach

In order to address the challenges posed by caregiving responsibilities, burdensome scheduling practices, and wage discrimination, we propose a number of reforms. The details of these proposed reforms are motivated in part by the evidence regarding their effects on women's labor market outcomes; much of that evidence has been generated by the experiences of state and local communities.

Paid Family Leave

Paid family leave benefits families and the overall economy (see, e.g., Baum and Ruhm 2013; Blau and Kahn 2013; Houser and Vartanian 2012; Rossin-Slater, Ruhm, and Waldfogel 2013). In light of the challenges facing workers who experience personal illness or caregiving demands—and given the importance of retaining those workers in the labor market—policies regarding paid leave must:

- Cover the range of family and medical needs that require time away from work;
- Be available to all workers, men and women equally;
- Provide adequate length of leave to address care needs; and
- Have a sufficiently high wage replacement rate to make a difference in people's lives.

As federal policymakers consider their options, they can learn from the experiences of three states—California, New Jersey, and Rhode Island—that have enacted statewide paid leave programs. These states provide important lessons about how to create successful paid family leave policies at the national level.

Cover the Range of Family and Medical Needs That Require Time Away from Work

An effective paid family and medical leave proposal must cover all the major reasons people need time away from their work. Family and medical leave is not exclusively about parental leave. As the population ages and women's labor force participation increases, more workers need time off to address either their own illness or that of a family member. Excluding any of these reasons from a paid leave policy would miss an opportunity to support both families' economic security and their labor force participation.

Be Available to All Workers, Men and Women Equally

Paid leave should cover all workers regardless of employer identity or size, or the worker's full-time or part-time status. It should also use an inclusive definition of family.

An effective paid leave program should be available to all workers, including those who are self-employed and those who work for small businesses. Placing boundaries on the availability of mandated paid leave negatively affects the labor market opportunities available to employed caregivers and others who require leave.

Paid leave also should be gender neutral, following the example of the Family and Medical Leave Act in providing eligible men and women with the same amount of leave. A mother-only policy assumes that only women do caregiving—in fact, women are breadwinners in 40 percent of families and men are taking on a growing share of the caregiving and other domestic responsibilities at home (Wang, Parker, and Taylor 2013).

Provide Adequate Length of Leave to Address Care Needs

Paid leave should entail at least 12 weeks of leave, allowing families enough time to deal with a serious illness or to care for a new child. Although 12 weeks falls short of the one year of parental care thought to ensure the best outcome for infants' development (and the six months of leave that is ideal for mothers' physical and mental health), it is more generous than the leave currently available, and will provide important benefits for parents and children alike (Schulte et al. 2017). It is also consistent with the level of generosity provided by states that have implemented paid leave programs, giving federal policymakers a better sense of how the proposal would work (National Conference of State Legislatures 2016). Furthermore, a 12-week leave allowance means that children born to two-parent families will have up to 24 weeks—or six months—of parental care if both parents use their full 12 weeks and schedule their leave periods sequentially.

Have a Sufficiently High Wage Replacement Rate to Make a Difference in People's Lives

Wages should be replaced at a level sufficient to protect families at a time when household expenses rise. A national paid leave program must replace enough of workers' wages to be economically meaningful and keep families afloat when they need time off for caregiving or for their own illness. It is important to provide a sufficient wage replacement rate considering that having a new child in the home, coping with a personal illness, or caring for a loved one often requires employees to cut back time spent at work during a period in which household expenses often rise.

Providing an economically meaningful replacement rate can also have benefits for businesses. A study of California's paid leave program found that employers in that state experienced greater worker retention following the enactment of paid leave, especially among those who employ low-income workers (Appelbaum and Milkman 2011; Horowitz et al. 2017).

Relatively generous wage replacement will also produce benefits in the form of reduced reliance on government benefits. In Rhode Island, where paid leave provides wage replacement between 55 percent and 66 percent, reduced use of government assistance was observed after paid leave was enacted (Houser and Vartanian 2012). Robust wage replacement plays an important role in realizing these benefits: Bernal and Fruttero (2008) found that, compared with unpaid leave, paid parental leave had a much bigger impact on long-term household incomes and labor participation for men and women alike.

National paid leave should therefore mimic New Jersey's 66 percent wage replacement, but with a cap that prevents benefits from being overly generous to high-income families. Wage replacement below this level would increase the likelihood that low-wage workers experience substantial economic hardship. This detail is consistent with Christopher Ruhm's Hamilton Project proposal. Ruhm recommends a replacement rate of 75 percent for low-wage workers, up to a ceiling of $1,323 per week.

Fair Scheduling

To ensure better economic outcomes for women, policies must address the way work and home lives are intertwined. Many individuals struggle to obtain enough work hours to make ends meet while also lacking the control over their schedules that would help them address their other obligations. Federal policymakers should ensure that workers can create boundaries between time for work and time for everything else by imposing fair scheduling. Policies to promote fair scheduling should:

- Require employers to bear costs associated with their last-minute decisions;
- Mitigate involuntary overwork and underwork; and
- Give workers the right to talk to their employer about flexible schedules without fear of reprisal.

Require Employers to Bear Costs Associated with Their Last-Minute Decisions

Employers should be required to provide advance notice of schedules, predictability pay for last-minute schedule changes, and reporting pay for shortened or on-call shifts to ensure that employees are able to balance their out-of-work responsibilities.

To address the unpredictable schedules faced by millions of workers, a national policy must ensure that workers have advance notice of their upcoming work schedule and relieve workers of the burden of last-minute scheduling changes when employers deem them necessary. This would be accomplished by requiring businesses to provide predictability pay when they alter a worker's schedule with less than seven days of notice. Workers would receive one hour of pay for each scheduling change made with less than seven days of notice. In addition, businesses would provide reporting pay in the form of two to four hours of wages when a shift is cancelled less than 24 hours in advance, as is required in San Francisco (American Legal Publishing n.d.) and Seattle (Municipal Code Corporation n.d.).

Mitigate Involuntary Overwork and Underwork

The Fair Labor Standards Act's overtime income threshold should be raised to further deter employers from requiring their employees to work long hours. A complementary policy to address excessive employer reliance on part-time workers would be to require employers to offer additional work hours to qualified part-time employees before hiring new employees.

The Fair Labor Standards Act provides some protection against overwork, increasing the cost for employers when they require certain employees to work more than 40 hours a week. While the law was intended to cover all hourly employees and a large share of salaried employees—excluding only those with earnings above a threshold—that threshold has not been significantly updated since 1975. To start, legislators should update the earnings threshold to keep pace with inflation. Second, with more workers being categorized as exempt from the overtime rule, policymakers need to consider whether the current definitions fit the modern workplace and provide sufficient worker protections.

The Obama administration updated these overtime policies in 2015, raising the overtime earnings threshold to $47,476, just below the inflation-adjusted 1975 level. In November 2015, however, a federal judge issued a temporary injunction blocking implementation of the reform. In blocking the rule, the Federal district court judge questioned the legality of any salary threshold.

The Trump administration's Department of Labor dropped the defense of the rule, but they did ask the 5th Circuit to approve the use of some (probably lower) threshold, arguing that there was "no basis to call into question a regulatory test that has been in place since the FLSA's inception." The appeal is tied to the department's plan to move forward with a lower threshold, possibly in the $32,000 to $35,000 range. EPI finds that this could leave 9.1 million fewer workers without overtime pay. (EPI 2017)

According to employer surveys, 50 percent or more of national companies have already implemented higher overtime standards and adjusted pay scales. This shows that the Obama overtime expansion would not be overly burdensome for businesses. Furthermore, New York and California have also acted to raise their OT standards even higher than the $47,476 level.

Give Workers the Right to Talk to Their Employer About Flexible Schedules Without Fear of Reprisal

Workers should have the right to negotiate work schedules with their employers without fear of reprisal, and require that employers listen and act where possible. Work schedules are an important concern for employees and their families, yet many U.S. workers are subject to disciplinary action or retaliation when asking for schedule changes. Union representation provides routes to engage in a conversation with employers about schedules, but with about 7 percent of private sector workers covered by a union contract, the large majority of workers need additional protection (BLS 2016a).

A right-to-request law establishes a process that gives employees the right to discuss their schedules or ask about scheduling flexibility without fear of negative consequences. Employees could ask to adjust their start or end times, switch a shift around, or even work remotely one day a week. Employers do not have to grant the request if it imposes undue hardship, but the right-to-request law requires that they have a compelling business reason for denying a request.

There is some evidence that this kind of policy improves labor market outcomes. Research on right-to-request laws in Australia, Germany, the Netherlands, New Zealand, and the United Kingdom shows that these policies are effective in limiting workers' work-life conflicts (Hegewisch and Gornick 2011; Lyness et al. 2012). These studies are not completely transferable to the United States, however, because these countries all have greater union coverage, which can help workers navigate a request process with their supervisor. In the United States most workers would have to learn about the law on their own and feel comfortable enough with their supervisor to take advantage of it. San Francisco and Vermont have recently passed and implemented right-to-request laws, but there is no research fully evaluating the effects of this legislation (Ludden 2014).

Combatting Wage Discrimination

To ensure equal pay for women, policy must combat wage discrimination. Three principles that redefine the power of knowledge about pay should be front and center in this effort:

- Prohibit employers from inquiring about a worker's salary history during the interview and hiring process;
- Ensure workers have the right to discuss pay; and

- Require employers to adopt pay transparency practices.

Prohibit Employers from Inquiring About a Worker's Salary History During the Interview and Hiring Process

Employers should be prohibited from asking about salary history during the interview or hiring process and relying on that information to set compensation. Federal lawmakers should consider the example set by Massachusetts, and since followed by several cities and states, in passing a measure banning employers from asking about salary histories during the job application process (Cunningham 2017). The state and local policies prohibit employers from screening job applications based on salary history, relying on past compensation to set pay, and asking workers about their salary history, including benefits and other compensation. Employers can confirm a prospective employee's compensation history, but only after an employment offer and compensation terms have been negotiated and extended (Cowley 2016; McGovern Tornone 2017; National Law Review 2017).

Ensure Workers Have the Right to Discuss Pay

Legislation should ban and create penalties sufficient to deter employers from retaliating against workers for discussing pay with their colleagues. We propose that all workers, including managers and supervisors, be included in a blanket prohibition of employer retaliation. The federal Paycheck Fairness Act, introduced first in 1997 and again this year, includes a provision that protects workers who disclose their pay to their colleagues. While this bill has not passed, policymakers can look to other examples: the Obama administration's 2014 executive order that banned federal contractors from retaliating against employees and job applicants "because such employee or applicant has inquired about, discussed, or disclosed the compensation of the employee or applicant or another employee or applicant" (White House 2014). Certain states have also passed laws addressing pay secrecy, differing in terms of what employees are covered and in which instances. Some states, for example, exclude public sector workers or managers and supervisors. Other states cover all workers, but only if those employees have instigated unequal pay claims (Kim 2015).

Require Firms to Adopt Pay Transparency Practices

Policy should incentivize employers to make disclosures of pay ranges and pay practices to employees and the government. U.S. employers with 100 or more employees are currently required to file an Employer Information Report EEO-1, or EEO-1 report, which provides a demographic breakdown of their workforce. The EEO-1 was updated in 2016 to require the separate reporting of pay data by sex, race, ethnicity, and job categories (U.S. Equal Employment Opportunity Commission [EEOC] n.d.).

Beginning in the spring of 2018, businesses with 100 or more employees would be required to add salary information to their EEO-1 reporting. The Trump administration, however, issued a "review and stay" of the rule in August 2017, meaning that businesses will not be required to report salary information. This serves as a substantial setback to efforts aimed at lessening pay inequality. The rule would have improved the EEOC's ability to investigate and address pay discrimination with individual employers and across industries and regions, encouraged employers to conduct voluntary pay audits, and provided some insight into the wage gap for employers that have not yet conducted pay audits. While the data gathered would have been limited and not made widely available to employees, it would serve as an important first step in the effort to better identify and end pay discrimination.

Prohibiting employer retaliation against workers who discuss pay is also not sufficient. Underpaid workers must still talk to their colleagues and raise the issue with their supervisor. This is often unlikely due to the taboo against salary discussions (Bierman and Gely 2004; Colella et al. 2007). Policymakers should therefore encourage employers to make affirmative disclosures of pay ranges and pay practices to employers and the government.

Many legal scholars have called for this kind of pay transparency to be mandatory, with University of Maryland School of Law's Deborah Thompson Eisenberg arguing for it on the grounds that pay discrimination is a "market failure caused by insufficient and asymmetric information about the value of work" (Eisenberg 2011, 951). Requiring companies to report pay information would be a further step toward ensuring that firms are properly valuing and rewarding employees, leading to a more efficient labor market (Eisenberg 2011).

Yet another approach has been proposed in Iceland, where recently introduced legislation would require employers to conduct audits on whether men and women are being paid fairly on a regular basis, and would impose fines on companies that do not take steps to ensure men and women are paid equally (Alderman 2017).

Conclusion

The proposed federal policies detailed in this paper would go a long way toward improving outcomes for women and all workers, thereby boosting the economy as well. Importantly, the policies must contain provisions for strict enforcement. Many labor laws rely on workers themselves to report violations, and private lawsuits are much more common than government investigations. This bottom-up enforcement is often insufficient given that many workers have well-founded fears about retaliation and are not willing to participate (Alexander and Prasad 2014).

The Obama administration outlined an agenda to improve labor law enforcement, which included a top-down approach of reaching out to industries or regions in which violations frequently occur, improving deterrence in those sectors, and clarifying boundaries of employment responsibility. The administration also ramped up outreach efforts around compliance and workers' rights and increased the number of investigators. And the EEOC's finalized initiative to collect pay data by race and gender will allow the EEOC to determine whether there are certain pay patterns for an employer, industry, or geographic region and potentially reveal where there is the need for enhanced scrutiny (U.S. EEOC 2016). The Trump administration is paring back these efforts (Meier and Ivory 2017).

In addition to strong enforcement, publicity and outreach campaigns are vital to the success of the proposals detailed in this paper. Evidence from state and local policies suggests that large groups of the population are unaware of worker protections, reducing their effectiveness (Appelbaum and Milkman 2011). Ensuring that these policies reach those they are intended to benefit is essential to producing better outcomes.

References

Alderman, Liz. 2017. "Equal Pay for Men and Women? Iceland Wants Employers to Prove It." *New York Times*, March 28.

Alexander, Charlotte S., and Arthi Prasad. 2014. "Bottom-Up Workplace Law Enforcement: An Empirical Analysis." *Indiana Law Journal* 89: 1069–1131.

American Legal Publishing Corporation. n.d. "Article 33G Section 3300G.4. San Francisco Police Code: Advance Notice of Work Schedules and Changes in Work Schedules." American Legal Publishing Corporation, Cincinnati, OH.

Appelbaum, Eileen, and Ruth Milkman. 2011, January. "Leaves That Pay: Employer and Worker Experiences with Paid Family Leave in California." Center for Economic and Policy Research, Washington, DC.

Appelbaum, Eileen, Heather Boushey, and John Schmitt. 2014, April. "The Economic Importance of Women's Rising Hours of Work." Center for American Progress and Center for Economic and Policy Research, Washington, DC.

Arnalds, Ásdís A., Guðný Björk Eydal, and Ingólfur V. Gíslason. 2013. "Equal Rights to Paid Parental Leave and Caring Fathers: The Case of Iceland." *Icelandic Review of Politics and Administration* 9 (2): 323–44.

Barach, Moshe A., and John J. Horton. 2017, March/April. "How Do Employers Use Compensation History?: Evidence from a Field Experiment." Social Science Research Network Scholarly Paper, Rochester, NY. https://papers.ssrn.com/abstract=2943034.

Baum, Charles L., and Christopher J. Ruhm. 2013, December. "The Effects of Paid Family Leave in California on Labor Market Outcomes." Working Paper No. 19741, National Bureau of Economic Research, Cambridge, MA.

Belogolovsky, Elena, and Peter A. Bamberger. 2014. "Signaling in Secret: Pay for Performance and the Incentive and Sorting Effects of Pay Secrecy." *Academy of Management Journal* 57 (6): 1706–33.

Belogolovsky, Elena, Peter A. Bamberger, Valeria Alterman, and David T. Wagner. 2016, January. "Looking for Assistance in the Dark: Pay Secrecy, Expertise Perceptions, and Efficacious Help Seeking Among Members of Newly Formed Virtual Work Groups." *Journal of Business and Psychology* 31 (4): 459–77.

Bernal, Raquel, and Anna Fruttero. 2008, November. "The Effect of Maternal Employment and Child Care on Children's Cognitive Development." *International Economic Review* 49 (4): 1173–1209.

Bertrand, Marianne and Sendhil Mullainathan. 2004. "Are Emily and Greg More Employable Than Lakisha and Jamal? A field Experiment on Labor Market Discrimination." *American Economic Review* 94 (4): 991-1013.

Bierman, Leonard, and Rafael Gely. 2004. "Love, Sex, and Politics—Sure. Salary—No Way: Workplace Social Norms and the Law." *Berkeley Journal of Employment and Labor Law* 25 (1): 167–91.

Blau, Francine D., and Lawrence M. Kahn. 2013. "Female Labor Supply: Why Is the United States Falling Behind?" *American Economic Review* 103 (3): 251–56.

Blau, Francine D., and Lawrence M. Kahn. 2016, January. "The Gender Wage Gap: Extent, Trends, and Explanations." Working Paper No. 21913, National Bureau of Economic Research, Cambridge, MA.

Boushey, Heather. 2016. *Finding Time: The Economics of Work–Life Conflict.* Cambridge, MA; London: Harvard University Press.

Boushey, Heather, and Bridget Ansel. 2016a, May. "Overworked America: The Economic Causes and Consequences of Long Work Hours." Washington Center for Equitable Growth, Washington, DC.

Boushey, Heather, and Bridget Ansel. 2016b. "Working by the Hour: The Economic Consequences of Unpredictable Scheduling Practices." Washington Center for Equitable Growth, Washington, DC.

Boushey, Heather, and Kavya Vaghul. 2016, April. "Women Have Made the Difference for Family Economic Security." Washington Center for Equitable Growth, Washington, DC.

Bowles, Hannah Riley, Linda Babcock, and Lei Lai. 2007, May. "Social Incentives for Gender Differences in the Propensity to Initiate Negotiations: Sometimes It Does Hurt to Ask." Organizational Behavior and Human Decision Processes 103 (1): 84–103.

Bureau of Labor Statistics (BLS). 2007–17. "Current Population Survey." Bureau of Labor Statistics, U.S. Department of Labor, Washington, DC.

Bureau of Labor Statistics (BLS). 2016. "National Compensation Survey: Employee Benefits in the United States, March 2016." Bureau of Labor Statistics, U.S. Department of Labor, Washington, DC.

Bureau of Labor Statistics (BLS). 2016. "Union Members — 2016." News Release, Bureau of Labor Statistics, U.S. Department of Labor, Washington, DC.

Cha, Youngjoo, and Kim A. Weeden. 2014, April. "Overwork and the Slow Convergence in the Gender Gap in Wages." *American Sociological Review* 79 (3): 457–84.

Colella, Adrienne, Ramona L. Paetzold, Asghar Zardkoohi, and Michael J. Wesson. 2007. "Exposing Pay Secrecy." *Academy of Management Review* 32 (1): 55–71.

Council of Economic Advisers (CEA). 2014, June. "The Economics of Paid and Unpaid Leave." Council of Economic Advisers, White House, Washington, DC.

Cowley, Stacy. 2016. "Illegal in Massachusetts: Asking Your Salary in a Job Interview." *New York Times*, August 2.

Cunningham, Larae Idleman. 2017, June. "Coming Soon to Your State? Laws Banning Inquiries into Applicant Salary History Gaining Steam." *HR Legalist* online.

DeLauro, Rosa. 2017. "Paycheck Fairness Act." H.R.1869, 115th Congress (2017–18). https://www.congress.gov/bill/115th-congress/house-bill/1869/text.

Earle, Alison, John Z. Ayanian, and Jody Heymann. 2006, May. "Work Resumption after Newly Diagnosed Coronary Heart Disease: Findings on the Importance of Paid Leave." *Journal of Women's Health* 15 (4): 430–41.

Economic Policy Institute. 2017. "Women's Economic Agenda: Creating an Economy That Works for Everyone." Economic Policy Institute, Washington, DC.

Eisenberg, Deborah Thompson. 2011. "Money, Sex, and Sunshine: A Market-Based Approach to Pay Discrimination." *Arizona State Law Journal* 43: 951–1020.

Elvira, Marta M., and Ishak Saporta. 2001, November. "How Does Collective Bargaining Affect the Gender Pay Gap?" *Work and Occupations* 28 (4): 469–90.

Fair Labor Standards Act of 1938 (FLSA) 29 U.S.C. 201, et seq. (1938).

Family and Medical Leave Act of 1993 (FMLA), Pub. L. 103-3, 107 Stat. 6 (1993).

Feinberg, Lynn Friss, and Rita Choula. 2012, October. "Understanding the Impact of Family Caregiving on Work." Fact Sheet 271, AARP Public Policy Institute, Washington, DC.

Frye, Jocelyn. 2016, November. "Refreshing the Conversation: Understanding Work-Family Challenges Facing Women of Color." Blog. U.S. Department of Labor, Washington, DC.

Fryer, Roland G., Devah Pager, and Jörg L. Spenkuch. 2011. "Racial Disparities in Job Finding and Offered Wages." NBER Working Paper 17642, National Bureau of Economic Research, Cambridge, MA, http://www.nber.org/papers/w17462.

Gely, Rafael, and Leonard Bierman. 2003. "Pay Secrecy/Confidentiality Rules and the National Labor Relations Act." *University of Pennsylvania Journal of Labor and Employment Law* 6 (1): 121–56. http://scholarship.law.missouri.edu/facpubs/219.

Glynn, Sarah Jane. 2014, June. "Breadwinning Mothers, Then and Now." Center for American Progress, Washington, DC.

Golden, Lonnie. 2015, April. "Irregular Work Scheduling and Its Consequences." Briefing Paper No. 394, Economic Policy Institute, Washington, DC.

Goldin, Claudia. 2014. "A Grand Gender Convergence: Its Last Chapter." *American Economic Review* 104 (4): 1091–119.

Goldin, Claudia, and Joshua Mitchell. 2017. "The New Life Cycle of Women's Employment: Disappearing Humps, Sagging Middles, Expanding Tops." *Journal of Economic Perspectives* 31 (1, Winter): 161–82.

GovDocs. 2017, February. "San Jose Opportunity to Work Ordinance Effective March 13." GovDocs, Saint Paul, MN.

Hall, Robert, and Alan Krueger. 2012. "Evidence on the Incidence of Wage Posting, Wage Bargaining, and On-the-Job Search." *American Economic Journal: Macroeconomics* 4 (4): 56–67.

Heckman, James J., and Dimitriy Masterov. 2007. "The Productivity Argument for Investing in Young Children." *Review of Agricultural Economics* 29: 446–93.

Hegewisch, Ariane, and Janet C. Gornick. 2011, May. "The Impact of Work-Family Policies on Women's Employment: A Review of Research from OECD Countries." *Community, Work & Family* 14 (2): 119–38.

Horowitz, Juliana Menasce, Kim Parker, Nikki Graf, and Gretchen Livingston. 2017, March. "Americans Widely Support Paid Family and Medical Leave, but Differ Over Specific Policies." Social & Demographic Trends Project, Pew Research Center, Washington, DC.

Houser, Linda, and Thomas P. Vartanian. 2012, January. "Pay Matters: The Positive Economic Impacts of Paid Family Leave for Families, Businesses and the Public." Center for Women and Work, New Brunswick, NJ.

Huet-Vaughn, Emiliano. 2015, August. "Do Social Comparisons Motivate Workers? A Field Experiment on Relative Earnings, Labor Supply and the Inhibitory Effect of Pay Inequality." Working Paper, Middlebury College, Middlebury, VT.

Institute for Women's Policy Research. 2014 January. "Quick Figures: Pay Secrecy and Wage Discrimination." IWPR No. Q016, Institute for Women's Policy Research, Washington, DC.

Jacobs, Jerry, and Kathleen Gerson. 2005. *The Time Divide: Work, Family, and Gender Inequality*. Cambridge, MA; London: Harvard University Press.

Kim, Marlene. "Pay Secrecy and the Gender Wage Gap in the United States." *Industrial Relations: A Journal of Economy and Society* 54, no. 4 (October 1, 2015): 648–67. doi:10.1111/irel.12109.

Landry, Bart. 2000. *Black Working Wives: Pioneers of the American Family Revolution.* Berkeley: University of California Press.

Lerner, Sharon. 2015, August. "The Real War on Families: Why the U.S. Needs Paid Leave Now." *In These Times.* http://inthesetimes.com/article/18151/the-real-war-on-families.

Lilly Ledbetter Fair Pay Act of 2009 Pub. L. 111–2, S. 181 (2009).

Ludden, Jennifer. 2014, April. "If You Want Flextime but Are Afraid to Ask, Consider Moving." NPR.org, Washington, DC.

Lyness, Karen S., Janet C. Gornick, Pamela Stone, and Angela R. Grotto. 2012, December. "It's All about Control: Worker Control over Schedule and Hours in Cross-National Context." *American Sociological Association* 77 (6): 1023–49.

Mas, Alexandre. 2014, October. "Does Transparency Lead to Pay Compression?" Working Paper No. 20558, National Bureau of Economic Research, Cambridge, MA.

McGovern Tornone, Kate. 2017, June. "Oregon Joins Growing List of States Banning Salary History Questions." HR Daily Advisor. https://hrdailyadvisor.blr.com/2017/06/13/oregon-joins-growing-list-states-banning-salary-history-questions.

Meier, Barry, and Danielle Ivory. "Under Trump, Worker Protections Are Viewed With New Skepticism." *The New York Times*, June 5, 2017, sec. Business Day. https://www.nytimes.com/2017/06/05/business/under-trump-worker-protections-are-viewed-with-new-skepticism.html.

Milli, Jessica, Yixuan Huang, Heidi Hartmann, and Jeff Hayes. 2017, April. "The Impact of Equal Pay on Poverty and the Economy." Institute for Women's Policy Research, Washington, DC.

Mitchell, David. 2017, February. "Stable and Predictable Scheduling as Antidote to Income Volatility." Aspen Institute, Washington, DC.

Moss-Racusin, Corinne, John F. Dovidio, Victoria L. Brescoli, Mark J. Graham, and Jo Handelsman. 2012. "Science Faculty's Subtle Gender Biases Favor Male Students." Proceedings of the National Academy of Sciences of the United States of America.

Municipal Code Corporation. n.d. "Title 14 Section 22. Seattle, Washington–Municipal Code: Secure Scheduling." Municipal Code Corporation, Tallahassee, FL. https://library.municode.com/wa/seattle/codes/municipal_code?nodeId=TIT14HURI_14.22SESC.

National Center for Education Statistics. 2016. "Digest of Education Statistics, table 303.10." National Center for Education Statistics, U.S. Department of Education, Washington, DC.

National Conference of State Legislatures. 2016. "State Family and Medical Leave Laws." National Conference of State Legislatures, Washington, DC.

National Labor Relations Act (NLRA) of 1935, Pub. L. 74-198, 49 Stat. 449 (1935).

The National Law Review. 2017. "Delaware Bans Compensation History Inquiries." *The National Law Review*, Western Springs, IL.

National Women's Law Center. 2013, January. "Fact Sheet: Lilly Ledbetter Fair Pay Act." National Women's Law Center, Washington, DC.

Patnaik, Ankita. 2015, April. "'Daddy's Home!' Increasing Men's Use of Paternity Leave." Briefing Paper, Council on Contemporary Families, Austin, TX.

Paycheck Fairness Act (S. 819 and H.R. 1869), introduced 1997 and 2017.

Rosenfeld, Jake, and Patrick Denice. 2015, October. "The Power of Transparency: Evidence from a British Workplace Survey." *American Sociological Review* 80 (5): 1045–68.

Rossin-Slater, Maya, Christopher J. Ruhm, and Jane Waldfogel. 2013, March. "The Effects of California's Paid Family Leave Program on Mothers' Leave-Taking and Subsequent Labor Market Outcomes: The Effects of California's PFL Program." *Journal of Policy Analysis and Management* 32 (2): 224–45.

Ruetschlin, Catherine, and Dedrick Asante-Muhammad. 2015. "The Retail Race Divide: How the Retail Industry Is Perpetuating Racial Inequality in the 21st Century." Demos and the NAACP, New York.

Sandstrom, Heather, and Sandra Huerta. 2013, September. "The Negative Effects of Instability on Child Development: A Research Synthesis." Low-Income Working Families Discussion Paper No. 3, Urban Institute, Washington, DC.

Schneider, Daniel, and Kristen Harknett. 2016, September. "Schedule Instability and Unpredictability and Worker and Family Health and Wellbeing." Working Paper 2016-09, Washington Center for Equitable Growth, Washington, DC.

Schneider, Jessica, and Elise Gould. 2016, July. "'Women's Work' and the Gender Pay Gap: How Discrimination, Societal Norms, and Other Forces Affect Women's Occupational Choices—and Their Pay." Economic Policy Institute, Washington, DC.

Schulte, Brigid, Alieza Durana, Brian Stout, and Jonathan Moyer. 2017, June. "Paid Family Leave: How Much Time Is Enough?" Better Life Lab, Washington, DC.

Stanczyk, Alexandra. 2016, October. "The Dynamics of Household Economic Circumstances around a Birth." Working Paper 2016-10, Washington Center for Equitable Growth, Washington, DC.

Stone, Pamela, and Lisa Ackerly Hernandez. 2013. "The All-or-Nothing Workplace: Flexibility Stigma and 'Opting Out' among Professional-Managerial Women." *Journal of Social Issues* 69 (2): 235–56.

Stone, Pamela, and Meg Lovejoy. 2004, November. "Fast-Track Women and the 'Choice' to Stay Home." *Annals of the American Academy of Political and Social Science* 596: 62–83.

Thomas, Mallika. 2016, September. "The Impact of Mandated Maternity Benefits on the Gender Differential in Promotions: Examining the Role of Adverse Selection." *Institute for Compensation Studies*, ILR School, Cornell University, Ithaca, NY.

U.S. Department of Labor. 2015, September. "The Cost of Doing Nothing: The Price We All Pay Without Paid Leave Policies to Support America's 21st Century Working Families." U.S. Department of Labor, Washington, DC.

U.S. Department of Labor .n.d. "Wage and Hour Division (WHD): Chapter 3. Use of the FMLA." U.S. Department of Labor, Washington, DC. https://www.dol.gov/whd/fmla/chapter3.htm.

U.S. Equal Employment Opportunity Commission (EEOC). 2016, September. "Press Release: EEOC to Collect Summary Pay Data." U.S. Equal Employment Opportunity Commission, Washington, DC.

U.S. Equal Employment Opportunity Commission (EEOC). n.d. "Small Business Fact Sheet: The Revised EEO-1 and Summary Pay Data." U.S. Equal Employment Opportunity Commission, Washington, DC.

Wang, Wendy, Kim Parker, and Paul Taylor. 2013, May. "Breadwinner Moms." Social & Demographic Trends Project, Pew Research Center, Washington, DC.

Warner, Judith. 2014, March. "The Women's Leadership Gap: Women's Leadership by the Numbers." Center for American Progress, Washington, DC.

Warren, Elizabeth. 2015, July. "Schedules That Work Act." S.1772, 114th Congress (2015–16). https://www.congress.gov/bill/114th-congress/senate-bill/1772.

White House. 2014, April 8. "Executive Order: Non-Retaliation for Disclosure of Compensation Information." Office of the Press Secretary, White House, Washington, DC.

World Economic Forum. 2013. "The Global Gender Gap Report 2013." World Economic Forum, Geneva.

VII. Skill Matchmaking in the Modern Economy

Building a Robust Apprenticeship System in the U.S. Why and How?

ROBERT I. LERMAN
Urban Institute

Skills drive productivity, competitiveness, and incomes. Economic growth is heavily dependent on the growth in human capital (Hanushek and Weissmann 2015). But what is human capital and what for that matter do we mean by "skills"? Too often, U.S. researchers have identified skills with two key measures: 1) academic attainment in terms of completion of schools and degrees; and 2) test scores on academic tests, usually tests of math and verbal capability. Any shortfall in these measures suggests the need for remedial action to help young people extend and complete schooling and to teach them better math and verbal capacities (Goldin and Katz 2008). This consensus view on skills is one reason spending on postsecondary education has grown rapidly and reached record levels per student. The National Center for Education Statistics reports that at the postsecondary level, the United States spent $27,900 per FTE student, 89 percent higher than the OECD average of $14,800.

Now, after mountains of student debt and enormous spending by federal, state, and local governments, the U.S. is said to face a serious skills mismatch in various occupations, especially those in technical fields. Some academics, consulting firms, and managers see weak skills of many American workers leading to skill shortages and limited economic growth (Deloitte 2011; Carnevale, Smith and Strohl 2010). One striking indication of a skills gap or mismatch is that German companies operating in the United States identify job skills as a key challenge to their success in the U.S. and encouraged the German Embassy to start a "Skills Initiative" to identify and share information about best practices in sustainable workforce development. Others assert that skills in the United States are not in short supply (Cappelli 2015; Osterman and Weaver 2014).

Unfortunately, debates on the adequacy of skills rarely incorporate an appropriately broad definition of skills. The virtual sole emphasis on academic skills as measured by math and verbal test scores and educational attainment is natural because that's where the data are. This emphasis fails to recognize that productivity depends at least as much on occupational competencies and employability skills, such as communication, teamwork, allocating resources, problem-solving, reliability and responsibility. The myriad nature of skills raises questions about whether added schooling and a targeted focus on academic test scores are the best ways of upgrading skills. So, too, does the recognition that many young people become disengaged from formal schooling, as reflected in weak high school outcomes and high dropout rates from community colleges.

Increasingly, policymakers and policy researchers are recognizing the need to shift from the "academic only" approach to teaching skills in schools. Instead, they see enormous potential in expanding apprenticeship, a model that combines work-based learning, production under a mentor/supervisor, wages, along with related courses. A wide body of evidence suggests that apprenticeships are far more cost effective in teaching skills, especially employability and occupational skills, than pure schooling. In Switzerland, perhaps the leading apprenticeship country, an astounding 95 percent of 25-year-olds have either a BA level degree or a recognized occupational certification, mainly through apprenticeship. About 70 percent of Swiss youth take up an apprenticeship, though some go on to university programs later.

Apprenticeship systems are one of the few mechanisms for improving both the supply and demand sides of the labor market. They are especially effective in teaching occupational and employability skills. Since classroom learning is applied quickly in real-world settings, workers are more likely to retain academic as well as occupational skills. Employers are more likely to create demanding, high productivity, and good-paying jobs when they can rely on those completing an apprenticeship to have mastered an array of relevant skills and to have gained experience in using those skills.

Apprenticeship expansion has become a bipartisan goal, endorsed and acted upon by President Trump at the beginning of his term and President Obama toward the end of his two terms. The Obama Administration allocated $175 million to 46 apprenticeship initiatives by nonprofits and community colleges. President Trump called for expanding apprenticeship at a White House ceremony last June. He endorsed a "moonshot" goal proposed by Salesforce CEO Marc Benioff to create 5 million apprenticeships in five years. Achieving 5 million apprenticeships would require a tenfold increase from today's 440,000 apprentices in civilian sectors and 95,000 in the military. Reaching the 5 million targets might sound impractical, but in fact, it would only require that the United States attain about the same share of apprentices in its workforce that Australia and England have already achieved. The president's first steps to achieve this goal were signing an executive order titled "Expanding Apprenticeship in America" and nearly doubling the funding for apprenticeships to $200 million.

Recognizing the need for fundamental reforms in the nation's apprenticeship system, the executive order establishes a task force to examine other administrative and legislative reforms, strategies for creating industry-recognized apprenticeships, and the best ways to encourage the private sector to create apprenticeships. The newly created task force is chaired by the Secretary of Labor and co-chaired by the Secretaries of Education and Commerce.[i]

Meanwhile, bipartisan bills in Congress call for providing tax credits to companies that offer apprenticeships. Several Republican and Democratic governors are taking steps aimed at expanding apprenticeships. For example, Republic Governor Scott Walker recently doubled the funding for Wisconsin's successful youth apprenticeship program. And Senator John Hickenlooper, a Colorado Democrat, is playing a leading role in creating a youth apprenticeship program modelled after the Swiss system.

This paper begins by defining apprenticeships and then discusses "why" apprenticeship can be an especially cost-effective approach to increasing skills, productivity, and ultimately wages. At this point, with the increasing acceptance of the rationale for expanding apprenticeship, the paper turns to the "how" questions. Is it feasible to scale up the U.S. system to reach numbers comparable to those in Australia and England while maintaining high quality? If so, what steps are required to do so?

Defining Apprenticeship and Explaining Its Advantages

Apprenticeship training is a highly developed system for raising the skills and productivity of workers in a wide range of occupations, with demonstrated success abroad and scattered examples of success domestically. Apprentices are employees who have formal agreements with employers to carry out a recognized program of work-based and classroom learning as well as a wage schedule that includes increases over the apprenticeship period. Apprenticeship prepares workers to master occupational skills and achieve career success. Under apprenticeship programs, individuals undertake productive work for their employer; earn a salary; receive training primarily through supervised, work-based learning; take academic instruction that is related to the apprenticeship occupation; and receive a certificate of completion. The programs generally last from two to four years. Apprenticeship helps workers to master not only relevant occupational skills but also other work-related skills, including communication, problem solving, allocating resources, and dealing with supervisors and a diverse set of co-workers. The course work is generally equivalent to at least one year of community college.

In Austria, Germany, and Switzerland, extensive apprenticeships offer a way of upgrading the quality of jobs, especially in manufacturing, commercial, and managerial positions.[ii] In these countries, apprenticeships begin mostly in the late high school years, absorbing 50-70% of young people on their way to valued occupational qualifications (Hoffman 2011). OECD reports (2009, 2010) highlight the role of a robust apprenticeship system in limiting youth unemployment.

Apprenticeships within the U.S. and elsewhere show how construction occupations can reach high wages and high productivity. The question is whether the model can be extended and attract firms to upgrade other occupations. Apprenticeship expansion holds the possibility of substantially improving skills and careers of a broad segment of the U.S. workforce. Completing apprenticeship training yields a recognized and valued credential attesting to mastery of skill required in the relevant occupation.

Apprenticeships are distinctive in enhancing both the worker supply side and the employer demand side of the labor market. On the supply side, the financial gains to apprenticeships are strikingly high. U.S. studies indicate that apprentices do not have to sacrifice earnings during their education and training and that their long-term earnings benefits exceed the gains they would have accumulated after graduating from community college (Hollenbeck 2008). The latest reports from the state of Washington show that the gains in earnings from various education and training programs far surpassed the gains to all other alternatives (Washington State Workforce Training and Education Coordinating Board 2014). A broad study of apprenticeship in 10 U.S. states also documents large and statistically significant earnings gains from participating in apprenticeship (Reed et al. 2012).

These results are consistent with many studies of apprenticeship training in Europe, showing high rates of return to workers. One recent study managed to overcome the obstacle that such studies tend to face where unmeasured attributes explain both who is selected for an apprenticeship and how well apprentices do in the labor market (Fersterer, Pischke, and Winter-Ebmer 2008); the authors did so by examining how an event unrelated to the apprenticeship (the firm staying in or going out of business) caused some apprentices to have full apprenticeships while others found their apprenticeships cut short. The estimates indicated that apprenticeship training raises wages by about 4% per year of apprenticeship training. For a three- to four-year apprenticeship, post-apprenticeship wages ended up 12-16% higher than they otherwise would be. Because the worker's costs of participating in an apprenticeship are often minimal, the Austrian study indicated high overall benefits relative to modest costs.

On the demand side, employers can feel comfortable upgrading their jobs, knowing that their apprenticeship programs will ensure an adequate supply of well-trained workers. Firms reap several advantages from their apprenticeship investments (Lerman 2014). They save significant sums in recruitment and training costs, reduced errors in placing employees, avoiding excessive costs when the demand for skilled workers cannot be quickly filled, and knowing that all employees are well versed with company procedures. Because employers achieve positive returns to their investments in apprenticeship, the worker and the government can save significantly relative to conventional education and training. After reviewing several empirical studies, Muehlmann and Wolter (2014) conclude that "...in a well-functioning apprenticeship training system, a large share of training firms can recoup their training investments by the end of the training period. As training firms often succeed in retaining the most suitable apprentices, offering apprenticeships is an attractive strategy to recruit their future skilled work force..." A recent detailed study conducted by the U.S. Department of Commerce and Case Western University (Helper et al. 2016) found that 40-50 percent returns to two expensive apprenticeship programs.

One benefit to firms rarely captured in studies is the positive impact of apprenticeships on innovation. Well-trained workers are more likely to understand the complexities of a firm's production processes and therefore identify and implement technological improvements, especially incremental innovations to improve existing products and processes. A study of German establishments documented this connection and found a clear relationship between the extent of in-company training and subsequent innovation (Bauernschuster, Falck, and Heblich 2009). Noneconomic outcomes are difficult to quantify, but evidence from Europe suggests that vocational education and training in general is linked to higher confidence and self-esteem, improved health, higher citizen participation, and higher job satisfaction (Cedefop 2011). These relationships hold even after controlling for income. An Australian study found that quality apprenticeships improve mental health (Buchanan 2016).

In the United States, evidence from surveys of more than 900 employers indicates that the overwhelming majority believe their programs are valuable and involve net gains (Lerman, Eyster, and Chambers 2009). Nearly all sponsors reported that the apprenticeship program helps them meet their skill demands—87% reported they would strongly recommend registered apprenticeships; an additional 11% recommended apprenticeships with some reservations. Other benefits of apprenticeships include reliably documenting appropriate skills, raising worker productivity, increasing worker morale, and reducing safety problems.

While apprenticeships offer a productivity-enhancing approach to reducing inequality and expanding opportunity, the numbers in the U.S. have declined in recent years to about one-tenth the levels in Australia, Canada, and Great Britain. Some believe the problems are inadequate information about and familiarity with apprenticeship, an inadequate infrastructure, and expectations that sufficient skills will emerge from community college programs. Others see the main problem as an unwillingness of U.S. companies to invest no matter how favorable government subsidy and marketing policies are. In considering these explanations, we should remember that even in countries with robust apprenticeship systems, only a minority of firms hires apprentices. Because applicants already far exceed the number of apprenticeship slots, the main problem today is to increase the number of apprenticeship openings that employers offer. Counseling young people about potential apprenticeships is a sensible complementary strategy to working with the companies, but encouraging interest in apprenticeship could be counterproductive without a major increase in apprenticeship slots.

Apprenticeships are a useful tool for enhancing youth development. Unlike the normal part-time jobs of high school and college students, apprenticeships integrate what young people learn on the job and in the classroom. Young people work with natural adult mentors who offer guidance but allow youth to make their own mistakes (Halpern 2009). Youth see themselves judged by the established standards of a discipline, including deadlines and the genuine constraints and unexpected difficulties that arise in the profession. Mentors and other supervisors not only teach young people occupational and employability skills but also offer encouragement and guidance, provide immediate feedback on performance, and impose discipline. In most apprenticeships, poor grades in related academic courses can force the apprentice to withdraw from the program. Unlike community colleges or high schools, where one counselor must guide hundreds of students, each mentor deals with only a few apprentices.

The high levels of apprenticeship activity in Australia, Great Britain, and Canada demonstrate that even companies in labor markets with few restrictions on hiring, firing, and wages are willing to invest in apprenticeship training. While no rigorous evidence is available about the apprenticeship's costs and benefits to U.S. employers, research in other countries indicates that employers gain financially from their apprenticeship investments (Lerman 2014).

Although apprenticeship training can prepare workers for a wide range of occupations, including engineering and architect[iii], apprenticeships are especially appropriate for skilled positions that do not require a B.A. degree.

Are Apprenticeship Skills Portable?

Concerns about whether the skills learned in apprenticeships bring the portability required to adapt to technical changes have recently surfaced (Hanushek, et al. 2017). Using cross-country regressions, the authors find countries that emphasize vocational education improve labor market outcomes in the short-run, but not in the long-run. While impacts are likely to vary by occupation, detailed studies indicate a high degree of skill portability associated with apprenticeship training.

To operationalize the concept of skill specificity, Geel and Gelner (2009) and Geel, Mure, and Gellner (2011) borrow an insight from Lazear (2009) that all skills are general in some sense, and occupation-specific skills are composed of various mixes of skills. The authors compile the key skills and their importance for nearly 80 occupations. They then use cluster analysis to estimate how skills are grouped within narrow occupations. This approach recognizes that skills ostensibly developed for one occupation can be useful in other occupations. It identifies occupational clusters that possess similar skill combinations within a given cluster and different skill combinations between clusters. Next, indices for each narrow occupation measure the extent to which the occupation is relatively portable between occupations within the same cluster and/or relatively portable between the initial occupation and all other occupations. The authors use these indices to determine how portability affects mobility, the wage gains and losses in moving between occupations, and the likelihood that employers will invest in training.

The authors test their hypotheses based on empirical analyses of German apprentices. One finding is that while only 42% of apprentices stay in their initial occupation, nearly two-thirds remain with either the occupation they learned as an apprentice or another occupation in the cluster using a similar mix of skills.

Second, those trained in occupations with more specific skill sets are most likely to remain in their initial occupation or move to occupations within the same cluster. Third, apprentices increase their wages when moving to another occupation within the same cluster but lose somewhat when moving to another cluster. Fourth, as Geel, Mure, and Gellner (2011) show, employers are especially likely to invest in apprenticeships with the most specific skill sets.

Other strong evidence of the high returns and transferability of German apprenticeship training comes from Clark and Fahr (2001). They examine the returns to apprenticeship for those who remain in the original apprentice occupation as well as losses that do or would occur from transferring to another occupation. The overall rates of return to each year of apprenticeship range from 8-12% for training in firms of 50 workers or more and from about 5.5-6.5% for firms of two to 49 workers. Transferring to another occupation can offset these gains, but the reduction is zero for those who quit and only 1.7% for those who are displaced from their job and shift to another occupation.

As found by Geel and Gellner (2009), the wage penalty varies with the distance from the original occupation. There is no penalty at all from displacement into a somewhat related occupation. Göggel and Zwick (2012) show the net gains or losses from switching employers and occupations differ by the original training occupation, with apprentices in industrial occupations experiencing wage advantages, while those in commerce, trading, and construction see modest losses. Finally, Clark and Fahr (2001) present workers' own views on their use of skills learned in apprenticeship training on their current jobs. Not surprisingly, 85% of workers remaining within their training occupation use many or very many of the skills they learned through apprenticeship. This group constitutes 55% of the sample. But, even among the remaining 45%, about two of five workers reported using many or very many of the skills from their apprenticeship and one in five used some of the skills. Overall, only 18% of all former apprentices stated they used few or no skills learned in their apprenticeships.

The findings show that the skills taught in German apprenticeship training are often general. Even when bundled for a specific occupation, the skills are portable across a cluster of occupations. Moreover, apprentices are quite likely to remain in occupations that use the skills they learned in their initial occupation. Apprenticeship skills do vary in terms of specificity and portability. But when the skills are less portable, firms are more likely to make the necessary investments and workers are less likely to change occupations significantly.

The general component of training is presumably stronger in school-based programs, because they are financed by government and/or individuals themselves. Yet, it is far from clear that these programs, especially the purely academic tracks in U.S. secondary schools and U.S. community colleges, offer more mobility. A high percentage of students drop out of both academic secondary and community college programs. Also, many of the community college programs are at least as specific as apprenticeship programs. Certificate programs within community colleges are almost entirely devoted to learning a narrow occupational skill, such as courses to become a phlebotomist, childcare assistant, or plastics-processing worker. Many U.S. school-based programs take place in for-profit colleges offering narrow programs, such as truck driving, medical assistant, and medical insurance billing and coding. Furthermore, skills often erode when they go unused. To the extent students learn general skills but rarely apply them and wind up forgetting them, their training is unlikely to offer upward mobility.

While community college and private for-profit students often take highly specific occupational courses, apprentices all take some general classroom courses. Thus, apprentice electricians learn the principles of science, especially those related to electricity. In most countries, collaboration takes place between public vocational schools and apprenticeship programs. In the U.S., apprentices often take their required "related instruction" in classes at community colleges or for-profit colleges (Lerman 2010). From this perspective, apprenticeship programs should be viewed as "dual" programs that combine work- and school-based learning, albeit with an emphasis on work-based learning.

Can the U.S. Scale Up Apprenticeships?

With the desirability of expanding apprenticeships gaining widespread support, the issue is now becoming one of feasibility. Can the U.S. scale apprenticeships and thereby widen the routes to rewarding careers and raising the quality and productivity of jobs? If so, how?

A common argument was that the U.S. lacked the cultural legacy of guilds common in parts of Europe, especially in Austria, Germany, and Switzerland. Another was that U.S. employers will never invest in the in-depth training of their workers. Both arguments have been weakened by experience. In the last two decades, Australia and England, two Anglo countries without the Continental European cultural legacy, have more than tripled their apprenticeships almost to the proportions of the labor force found in Austria, Germany, and Switzerland. In the U.S., South Carolina managed to increase the number of companies adopting apprenticeship programs from 90 in 2007 to over 800 in the subsequent eight years, using a combination modest funding and a high- quality marketing and sales initiative.

The biggest reason for lacking a robust apprenticeship system in the U.S. is the failure to try. Today, even after recent allocations demonstration funding, government spending on apprenticeships is minimal compared with spending by other countries as well as compared with what it costs to pay for less effective career and community college systems that provide education and training for specific occupations. While total government funding for apprenticeship in the U.S. has only been about $100 to $400 per apprentice annually, federal, state, and local government spending annually per participant in two-year public colleges was approximately $11,400 in 2008 dollars (Cellini 2012). Not only are government outlays sharply higher, but the cost differentials are even greater after accounting for fact that the foregone earnings of college students as they learn far exceed any forego earnings apprentices experience. Nearly all other countries with significant apprenticeship programs pay for the off-job courses required in an apprenticeship. The U.S. rarely does so.

Overall, the federal government has been spending less than $30 million annually to supervise, market, regulate, and publicize the system. Many states have only one employee working under their OA. Were the U.S. to spend what Britain spends annually on apprenticeship, adjusting for the differences in the labor force, it would provide at least $9 billion per year for apprenticeship. Note that the Federal Pell Grant program for low- and lower-middle income college students costs about $33 billion per year, with a good chuck of the spending going toward career-focused programs in community and career colleges. Thus, at least some of the low apprenticeship penetration to a lack of public effort in promoting and supporting apprenticeship and to heavy subsidies for alternatives to apprenticeship.

Still, other barriers to expansion are significant. One is limited information about apprenticeship. Because few employers offer apprenticeships, most employers are unlikely to hear about apprenticeships from other employers or from workers in other firms. Compounding the problem is both the difficulty of finding information about the content of existing programs and the fact that developing apprenticeships is complicated for most employers, often requiring technical assistance that is minimal in most of the country.

Another barrier is employer misperceptions that apprenticeship will bring in unions. There is no evidence that adopting an apprenticeship program will increase the likelihood of unionization, but reports about such close links persist. An additional barrier is the asymmetric treatment of government postsecondary funding, with courses in colleges receiving support and courses related to apprenticeship receiving little financial support. Policies to reduce the government spending differentials between college subsidies and apprenticeship subsidies can help overcome this barrier.

Whether to emphasize apprenticeships beginning in late high school or after high school involves tradeoffs. High school programs improve the likelihood of government funding for academic courses related to apprenticeships. Given the consensus that the government should fund students through secondary school, paying for the related instruction of high school apprentices becomes a nondiscretionary part of budgets. When apprentices are beyond high school, government funding for related instruction must come out of discretionary expenses. International experience demonstrates the feasibility of youth apprenticeships; youth can attain serious occupational competencies while completing secondary education.

Apprenticeships in the late teenage years improve the nonacademic skills of youth at a critical time. In countries with little or no youth apprenticeship, structured work experience is less common, limiting the ability of youth to develop critical employability skills such as teamwork, communication, problem solving, and responsibility. Early apprenticeships can help engage youth and build their identity (Halpern 2009). Apprentices work in disciplines that are interesting and new; they develop independence and self-confidence through their ability to perform difficult tasks. Youth try out new identities in an occupational arena and experience learning in the context of production and making things.

From an economic perspective, apprenticeships for youth can be less costly for employers. Wages can be lower partly because youth have fewer medium- and high-wage alternatives and partly because youth have fewer family responsibilities, allowing them to sacrifice current for future income more easily. While Swiss firms invest large amounts of dollars in their apprenticeship programs, they pay their young apprentices very low wages during the apprenticeship period. Another economic advantage is that starting earlier in one's career allows for a longer period of economic returns to training.

For the U.S., scaling apprenticeship in the last years of high school is difficult. The aversion to tracking students too early into an occupational sequence is a common objection to youth apprenticeship. Importantly, high school officials are generally averse to adding youth apprenticeship to their already extensive agenda, including implementing Common Core standards and school and teacher accountability standards as well as dealing with charter schools and vouchers. In the early 1990s, opposition to youth apprenticeship in the U.S. came from unions and others who worried about eroding the apprenticeship brand with less intensive training programs.

While the verdict is still out on whether the U.S. can achieve scale in apprenticeships, its best chance is to assess where the system needs to go and to take incremental steps to get there.

Ten Elements for a Robust Apprenticeship System

Broad political and industry support are necessary but not sufficient to build and sustain a robust apprenticeship system. In addition, several elements are required for the system to work well. These include:

- Effective branding and broad marketing
- Incentives for selling and organizing apprenticeships to private, public employers
- Programs to develop credible occupational standards with continuing research
- End-point assessments of apprentices & programs
- Certification body to issue credentials
- Making apprenticeships easy for employers to create and to track progress
- Funding for off-job classes quality instruction
- Counseling, screening prospective apprentices to insure they are well-prepared
- Training the trainers for apprenticeship
- Research, evaluation and dissemination

Branding Apprenticeship

Recent successes in Britain and South Carolina have been accompanied by a concerted effort to create apprenticeship as a distinctive brand. South Carolina chose to link apprenticeship with local pride with the brand name of "Apprenticeship Carolina". Britain began its growth with the name "Modern Apprenticeships" but subsequently allowed the apprenticeship label to stand on its own while copywriting the term. It is now illegal to call an employer training program an apprenticeship unless it is under the official apprenticeship system. At the same time, Britain spent millions of dollars advertising apprenticeships, including advertisements on the London subways.

Selling and Organizing Apprenticeships

Branding and broad marketing will not suffice without a well-developed system for selling and organizing apprenticeships. An employer convinced by an advertisement must have a place to call to learn about and implement an apprenticeship in the organization. Britain's success in expanding apprenticeships offers one example for how to create successful national and decentralized marketing initiatives. Alongside various national efforts, including the National Apprenticeship Service and industry skill sector councils, the British government provided funding for the off-job instruction in apprenticeships to private training organizations and to Further Education colleges. These funds were sufficient to encourage these organizations to sell and organize apprenticeships with employers. In fact, the British approach has buttressed an association of private companies that engage in the kind of retail marketing required to persuade employers to offer apprenticeships. Another step is the British government's initiative to create apprenticeships within the civil service, specifying that apprentices should constitute 2.3 percent of government employment.

The success of South Carolina in selling and organizing apprenticeships has depended on the skills of small staff built originally by Ann-Marie Stieritz, the director of Apprenticeship Carolina. She hired individuals who understand businesses, who are engaging, who had worked in companies, ideally the business services industry, and who knew how to develop and manage relationships. She did not require knowledge or experience of apprenticeship. For the first two weeks, the staff engaged in a total immersion learning process about apprenticeship, where they learned about the concept of apprenticeship, apprenticeship regulations and forms, and saw apprenticeship programs first hand. The staff worked closely with Ron Johnson, a career employee and the federal apprenticeship's office representative for South Carolina. The presence of Johnson and his flexibility in pushing for the approval of company programs was important in the initiative's ability to expand within the context of the registered apprenticeship system.

The expansion of apprenticeship has involved reaching out across broad industry sectors, including advanced manufacturing, health care, and information technology. Apprenticeship marketing often takes place in the context of state and local economic development efforts to attract new businesses. The program's work with companies on their training needs is marketed as a reason for a firm to locate in South Carolina. Workforce Innovation and Opportunity (WIOA) agencies are also cooperating, sometimes providing on-the-job training subsidies in the context of apprenticeship. The chamber of commerce publicizes apprenticeship through forums, newsletters and committee meetings. The value added by Apprenticeship Carolina comes mainly from the program's ability to work with business to diagnose their skill demands, including what they see as an ideal set of skills that they want workers to master.

Credible Occupational Standards

Nearly all countries with robust apprenticeship systems create occupational frameworks for apprenticeship that all employers training in the relevant occupation mainly follow, with modest additions relating to their own organization. The current US "registered apprenticeship" system is unique in requiring individual companies or other sponsors (such as unions) that wish to register their programs to supply their own skill frameworks and curriculum. In half of the states, the approval process is subject to the preferences of state agencies that are often highly restrictive and that require excessive numbers of journeymen/mentors (people who have completed an apprenticeship in the field or have occupational expertise developed elsewhere) per apprentice. Pennsylvania, for example, mandates a ratio of four to one.

The structure for registered apprenticeships in the U.S. leads to skill frameworks that are often uneven and highly variable. While joint employer-union construction apprenticeship programs generally use common frameworks for each occupation, even union programs can vary from state to state.

Employers on their own rarely have the time nor common vision across employers to develop frameworks on their own. Moreover, the frameworks should reflect the interests of the apprentices as well as the interests of the employers. This is especially the case if the public sector provides some funding for the programs to take account of the general skills (skills that have value outside the training firm) taught.

Countries vary in their approaches but all rely on the cooperation of the public and private sectors. The Institute for Apprenticeship in England recently began operating, with the responsibility to oversee skill frameworks initially created by leading employers using the occupation. In Switzerland, the Federal Office for Professional Education and Technology, together with cantons, employers, trade associations and unions, participate in framing the occupational standards for about 250 occupations (Hoeckel, Field and Grubb 2009). The canton vocational education programs implement and supervise the vocational schools, career guidance, and inspection of participating companies and industry training centers. Professional organizations develop qualifications and exams and help develop apprenticeship places. Occupational standards in

Germany are determined primarily by the "social partners", including government, employer, and employee representatives (Hoeckel and Schwartz 2009). The chambers of commerce advise participating companies, register apprenticeship contracts, examine the suitability of training firms and trainers, and set up and grade final exams.

The content of skill requirements in apprenticeships includes academic courses and structured work-based training. In each field, the requirements are to complete the coursework in a satisfactory manner and to demonstrate the apprentice's ability to master a range of tasks. In some systems, there are a set of general tasks that apply to a family of occupations (say, metalworking) and tasks that apply to a specific occupation (say, tool mechanics or metal construction and shipbuilding). While the tasks vary widely across occupations, all involve the application of concepts and academic competencies.

Under a contract from the U.S. Department of Labor, the Urban Institute in collaboration with the American Institute for Innovative Apprenticeship has begun publishing competency-based occupational frameworks for apprenticeships in several occupations.[iv] This approach could form the foundation for what President Trump's executive order calls "industry-recognized apprenticeships." The idea of moving away from the registered apprenticeship approach of recognizing and registering occupational apprenticeship programs on a company-by-company basis has been criticized loosening quality standards. The argument is that limiting the government role in approving programs will lower the quality of apprenticeships. In fact, developing broad-based industry standards is likely to raise quality and to move the US system toward national frameworks that are common in all other countries with robust apprenticeship programs.

Assessments and Certifications

The extent to which systems develop third-party assessments varies across countries. In Germany, each apprentice is subject to an examination by six to nine experts in the occupation, including representatives from Chambers of Commerce and educators. Several organizations in Britain provide what are called end-point assessments as well as interim assessments. One of the largest is City and Guilds, a private organization that provides curricula as well as assessment services for a large number of apprenticeships. In addition, most countries provide audits of overall programs, including the on-the-job learning and the quality of off-job related instruction.[v]

In the U.S., federal and state offices lack the staff to audit programs for quality or to provide third-party assessments of apprentices. State and federal apprenticeship agencies do award certifications of completion based on the reports by employers of the progress of apprentices through their programs. Although completion certificates under the registered apprenticeship system (both state and federal) are supposed to be portable throughout the U.S., not all states recognize completers from state or federal programs they view as subpar.

Making Apprenticeships Easy for Employers to Create

Marketing to firms through existing federal and state agencies has not worked to scale apprenticeships so far. Although the lack of staff and minimal funding for even the off-job components of apprenticeships play major roles, the system's complexity can also be a barrier. South Carolina's sales representatives show that it is possible in some contexts to simplify the process of developing an apprenticeship occupational framework and doing all the paper work necessary to register the program. The state apprenticeship tax credit of $1,000 per apprentice per year is also simple to claim. However, the case of South Carolina is an exception. One reason is that the absence of common occupational frameworks that are well-recognized as yielding quality outcomes. Another is the federal-state approval process. And a third is the absence of talented people who can sell and organize apprenticeships as they become human resource consultants.

Funding for Off-Job Classes Related to the Apprenticeships

One can make a strong theoretical and practical case for the training firm not funding the off-job learning in an apprenticeship. Theoretically, the skills learned in the off-job courses are general in the Becker sense that the added productivity of the worker can be applied not only to his or her current employer but to many other employers. For this reason, the employer cannot recoup the provision of this general training. The worker gains the benefit, but the government shares his or her gain in the form of higher taxes and reduced transfers. On the practical side, the government already funds a significant share of the costs of courses aimed at teaching occupational skills but does so in a way that is far less cost-efficient than apprenticeship.

Judging by the case of England, financing the cost of delivering courses for apprenticeships by training organizations could be enough to encourage them to sell employers on apprenticeships. Using a pay for performance model, technical education and training organizations would earn revenue only for apprenticeships that each college or organization stimulates.

The government could reap savings from this approach since every apprenticeship slot stimulated by an already funded college/training organization increases the work-based component of training borne by the employer and reduces the classroom-based component often borne by government. Consider the following example for community colleges. Assume the work-based component amounts to 75% of the apprentice's learning program and the school-based courses are only 25% of the normal load for students without an apprenticeship. By allowing training providers to keep more than 25% of a standard full-time-equivalent cost provided by federal, state, and local governments in return for providing the classroom component of apprenticeship, the community colleges and other training organizations would have a strong incentive to develop units to stimulate apprenticeships.

Another possibility is to emphasize apprenticeships in the context of existing high school-based career and technical education programs. Since high school CTE course are already financed as an entitlement, the funds to complement work-based learning in apprenticeships would be readily available. Good places to start are career academies—schools within high schools that have an industry or occupational focus—and regional career and technical education (CTE) centers. Over 7,000 career academies operate in the U.S. in fields ranging from health and finance to travel and construction (Kemple and Willner 2008). Career academies and CTE schools already include classroom-related instruction and sometimes work with employers to develop internships. Because a serious apprenticeship involves learning skills at the workplace at the employer's expense, these school-based programs would be able to reduce the costs of teachers relative to a full-time student. If, for example, a student spent two days per week in a paid apprenticeship or 40% of time otherwise spent in school, the school should be able to save perhaps 15% to 30% of the costs. Applying these funds to marketing, counseling, and oversight for youth apprenticeship should allow the academy or other school to stimulate employers to provide apprenticeship slots. Success in reaching employers will require talented, business-friendly staff who are well trained in business issues and apprenticeship.

Allowing the use of Pell grants to pay at least for the classroom portion of a registered apprenticeship program makes perfect sense as well. Currently, a large chunk of Pell grants pays for occupationally oriented programs at community colleges and for-profit career colleges. The returns on such investments are far lower than the returns to apprenticeship. The Department of Education already can authorize experiments under the federal student aid programs (Olinsky and Ayres 2013), allowing Pell grants for some students learning high-demand jobs as part of a certificate program. Extending the initiative to support related instruction (normally formal courses) in an apprenticeship could increase apprenticeship slots and reduce the amount the federal government would have to spend to support these individuals in full-time schooling.

The GI Bill already provides housing benefits and subsidizes wages for veterans in apprenticeships. However, funding for colleges and university expenses is far higher than for apprenticeship. Offering half of the GI Bill college benefits to employers hiring veterans into an apprenticeship program could be accomplished by amending the law. However, unless the liberalized uses of Pell grants and GI Bill benefits are linked with an extensive marketing campaign, the take-up by employers is likely to be limited.

Counseling, Screening Prospective Apprentices to Insure They Are Well-Prepared

Apprenticeships typically require apprentices and employers to commit to a long-term, 2-5-year training program. Before making any commitment of this duration, apprentices should have a clear understanding of the occupation they are entering, the production and learning activities they will undertake during the apprenticeship, and the long-term career opportunities that completing the apprenticeship will afford.

In the U.S., formal counseling does take place in high schools, usually during sophomore year, for those considering entering a youth apprenticeship program. But, typically U.S. workers enter registered apprenticeships well after high school in their mid-to-late 20s. Although some workers may receive counseling services from American Job Centers, most learn about apprenticeships informally, having bounced various occupations and jobs. They learn from media, friends and families about apprenticeship openings and apprenticeship occupations. The informal knowledge may not be enough for apprentices to appreciate fully what the job, career, and work atmosphere will entail. Still, unlike those going through a degree program before entering a profession, apprentices will learn about the occupation within the first few months of their education and training.

Typically, the screening process brings out information on the test scores in math and verbal, work experience, and some gauge of how enthusiastic apprentices are when applying to an employer. However, increasing opportunities for apprentices and employers to learn more about each other before an agreement is formalized should be on the agenda for expanding apprenticeships in the U.S. Improved systems for matching prospective apprentices with current and future apprenticeships offered by employers could improve this process.[vi]

Train the Trainers

The quality of trainers is an important element in the success of apprenticeships. That is one reason why several European systems devote considerable time to training and certifying trainer/mentors of apprenticeship. In the late 1990s, the European Centre for the Development of Vocational Education (Cedefop), decided to promote the sharing of best practices for training trainers and other vocational education instructors across 22 national networks.

In Germany, anyone who wishes to serve as a trainer in the apprenticeship system must demonstrate both technical qualifications and appropriate personal attributes. Trainers are skilled workers who have several years of professional experience and have taken a two-week course at a chamber of industry and commerce or chamber of crafts and trades to prepare for the AEVO exam. Trainer aptitude includes the ability to independently plan, conduct, and monitor vocational training, as well as to plan and prepare training programs, to collaborate in the hiring of apprentices, and to conduct and conclude training. Today, some 90,000 people per year take the trainer aptitude examination.

A trainer must be able to examine the capacity of the company to offer training in the desired certified trade; to create a company training program on the basis of a training regulation geared toward the job-specific work and business processes; to create the necessary conditions and foster a motivating learning environment; to select training methods and materials appropriate to the target group and to deploy them in specific situations; to support apprentices with learning difficulties through customized training design and counselling; to prepare apprentices for the final and journeyman examination; and finally to bring the training program to successful conclusion.

The U.S. lacks any formal system for insuring trainers of apprentices have the requisite skills and personal attributes to perform well.

Research, Evaluation, and Dissemination

An infrastructure for research, evaluation, dissemination, and peer support can play an important role in scaling up and continuous improving the apprenticeship system. Such functions offer clear externalities to workers and employers. The federal government should sponsor the development of a public/private partnership that houses an information clearinghouse, a peer support network, and a research and evaluation program on apprenticeship. Research could be conducted on the effectiveness of apprenticeships in insuring that workers learn the key occupational, employability, and academic skills, on the short-term and long-term impacts on earnings compared with other approaches to education and training, and on the regulatory aspects of apprenticeship. Also important are topics especially relevant to employers, such as the return to apprenticeship from the employer perspective and the net cost of sponsoring an apprentice after taking account of the apprentice's contribution to production. The evaluations should cover best practices for marketing apprenticeship, incorporating classroom and work-based learning by sector, and counseling potential apprentices.

An information clearinghouse can document international experience with apprenticeship, including skill frameworks for apprenticeships used in various countries. Finally, the public/private institute would engage in dissemination about the impacts of apprenticeships and best practices in apprenticeship.

Summing Up

Expanding apprenticeship is a potential game-changer for improving the lives of millions of Americans and preventing further erosion of the middle class. Apprenticeships widen routes to rewarding careers by upgrading skills, including occupational skills but also math, reading, and employability skills. Taking math, reading, and writing in the context of using these competencies in the workforce will increase the motivation of many workers and the efficacy of the delivery process. Given the ability of workers to learn more, remain well motivated, and notice how to make innovations at the workplace, firms will have an increased incentive to adopt "high road" strategies and make them work. Such an approach may be one of the only ways the firm can attract and sustain workers.

Yet, today, funding for the "academic only" approach to skill development dwarfs the very limited amounts available to market and support apprenticeship. Instead of spending well over $11,000 per year on students in community college career programs, why not shift resources toward far more cost-effective apprenticeship programs? Apprenticeship programs yield far higher and more immediate impacts on earnings than community or career college programs yet cost the student and government far less. Community college graduation rates, especially for low-income students, are dismally low. Even after graduating, individuals often have trouble finding a relevant job. For students in postsecondary education, foregone earnings are one of the highest costs and many incur considerable debt. In contrast, participants in apprenticeships rarely lose earnings and often earn more than if they did not enter an apprenticeship. Rarely must apprentices go into debt while they learn. And apprentices are already connected with an employer and can demonstrate the relevant credentials and work experience demanded by other employers. Another advantage is the net gains flowing to employers from apprenticeship programs.

Structural barriers require some up-front government investments to help build robust apprenticeship system in the U.S. Investments in marketing and standard development, along with ongoing support for the off-job costs of apprenticeship, are likely to attract large numbers of employers. As more employers adopt apprenticeship strategies successfully, network effects could well take over, with employers learning from each other about the value of apprenticeship. At some point, we may see a tipping point when government spending on marketing becomes far less necessary. Institutional change of this magnitude is difficult and will take time but will be worthwhile in increasing earnings of workers in middle-skill jobs, widening access to rewarding careers, enhancing occupational identity, increasing job satisfaction, and expanding the middle class.

It is past time for federal and state governments to make a genuine effort to build an extensive and high value apprenticeship system. Without such an effort, we are not likely to upgrade skills and jobs and we are likely to continue to expend vast resources on a college-based, academic-only system that fails millions of students. With such an effort, I believe U.S. employers will follow their counterparts in other countries, create a significant number of apprenticeship slots, and realize gains in recruitment, workforce quality, and improved productivity. Institutional change of this magnitude is difficult and will take time but will be worthwhile in increasing earnings of workers in middle-skill jobs, widening access to rewarding careers, raising national productivity, enhancing occupational identity, increasing job satisfaction, and expanding the middle class.

Endnotes

[i] The Secretary of Labor announced the task for on October 16. For the membership of the task force, see https://www.dol.gov/newsroom/releases/osec/osec20171016

[ii] For a list of occupations using apprenticeships in several countries, see the occupational standards section of the American Institute for Innovative Apprenticeship website at www.innovativeapprenticeship.org.

[iii] The United Kingdom features an array of apprenticeships with college degrees in a variety of fields. See https://www.instituteforapprenticeships.org/apprenticeship-standards/?levelFrom=5

[iv] See https://innovativeapprenticeship.org/us-apprenticeships/ for examples.

[v] In England, Ofsted, an agency that reports directly to the Parliament, rates the quality of apprenticeship providers.

[vi] For an example of apprenticeship matching site, see https://www.loveapprenticeship.com/ .

References

Bauernschuster, Stefan, Oliver Falck, and Stephan Heblich. 2009. "Training and Innovation." *Journal of Human Capital* 3 (4): 323-353.

Buchanan, John, Catherine Raffaele, Nick Glozier and Aran Kanagaratnam. 2016. *Beyond mentoring: social support structures for young Australian carpentry apprentices*. Adelaide, Australia: National Centre for Vocational Education Research. https://www.ncver.edu.au/publications/publications/all-publications/2865 [July 22, 2017].

Cappelli, Peter. 2015. "Skill Gaps, Skill Shortages and Skill Mismatches: Evidence and Arguments for the United States." *ILR Review*. March. 68: 251-290.

Carnevale, Anthony P., Nicole Smith, and Jeff Strohl, "Help Wanted: Projections of Jobs and Education Requirements through 2018," Center on Education and the Workforce, Georgetown University, 2010, pp. 1-5, http://cew.georgetown.edu/jobs2018. [September 16, 2017].

Clark, Damon, and René Fahr. 2001. The Promise of Workplace Training for Non-College-Bound Youth: Theory and Evidence from German Apprenticeship. IZA Discussion Paper 378. Bonn, Germany: IZA. http://www.iza.org/en/webcontent/publications/papers/viewAbstract?dp_id=378 [October 10, 2015].

Cedefop. 2011. *Vocational Education and Training Is Good for You: The Social Benefits of VET for Individuals.* Research Paper 17. Luxembourg: Publications Office of the European Union.

Cellini, Stephanie. 2012. "For-Profit Higher Education: An Assessment of Costs and Benefits." *National Tax Journal.* 65(1). 153-180.

Deloitte. 2011. "Boiling Point? The Skills Gap in U.S. Manufacturing." Manufacturing Institute. http://www.themanufacturinginstitute.org/~/media/A07730B2A798437D98501E798C2E13AA.ashx [September 14, 2015].

Fersterer, Josef, Jorn-Steffen Pischke, and Rudolf Winter-Ebmer. 2008. "Returns to Apprenticeship Training in Austria: Evidence from Failed Firms." *Scandinavian Journal of Economics*. 110 (4): 733-53.

Geel, Regula, and Uschi Backes-Gellner. 2009. Occupational Mobility Within and Between Skill Clusters: An Empirical Analysis Based on the Skill-Weights Approach, Working Paper No. 47. Swiss Leading House on Economics of Education, Firm Behavior and Training Policies. Zurich: Swiss Federal Office for Professional Education and Technology. http://ideas.repec.org/p/iso/educat/0047.html. [January 8, 2014].

Geel, Regula, Johannes Mure, and Uschi Backes-Gellner. 2011. "Specificity of Occupational Training and Occupational Mobility: An Empirical Study Based on Lazear's Skill-Weights Approach." *Education Economics* 19 (5): 519-535.

Göggel, Kathrin, and Thomas Zwick. 2012. "Heterogeneous Wage Effects of Apprenticeship Training." *Scandinavian Journal of Economics*. 114 (3): 756-79.

Goldin, Claudia, and Lawrence F. Katz. 2008. *The Race Between Education and Technology*. Cambridge, MA: Harvard University Press.

Halpern, Robert. 2009. *The Means to Grow Up. Reinventing Apprenticeship as a Developmental Support in Adolescence.* New York: Routledge.

Hanushek, Eric A., Guido Schwerdt, Ludger Woessmann and Lei Zhang. 2017. "General Education, Vocational Education, and Labor-Market Outcomes over the Lifecycle." *Journal of Human Resources*. Winter. vol. 52 (1) 48-87.

Hansson, Bo. 2007. "Company-based determinants of training and the impact of training on company performance." *Personnel Review*. 36(2), 311-331.

Helper, Susan, Ryan Noonan, Jessica Nicholson, and David Langdon. 2016. *The Benefits and Costs of Apprenticeship Programs: A Business Perspective*. Washington, DC: U.S. Department of Commerce.

Hollenbeck, Kevin. 2008. "State Use of Workforce System Net Impact Estimates and Rates of Return." Presented at the Association for Public Policy Analysis and Management (APPAM) Conference, Los Angeles. http://research.upjohn.org/confpapers/1 [September 4, 2012].

Hoffman, Nancy. 2011. *Schooling in the Workplace: How Six of the World's Best Vocational Education Systems Prepare Young People for Jobs and Life*. Cambridge, MA: Harvard Education Press.

Lazear, Edward. 2009. "Firm-Specific Human Capital: A Skill-Weights Approach." *Journal of Political Economy* 117: 914–40.

Lerman, Robert. 2014. "Do firms benefit from apprenticeship investments?" IZA World of Labor. http://wol.iza.org/articles/do-firms-benefit-from-apprenticeship-investments-1.pdf [September 16, 2017].

Lerman, Robert, Lauren Eyster, and Kate Chambers. 2009. "The Benefits and Challenges of Registered Apprenticeship: The Sponsors' Perspective." Washington, DC: U.S. Department of Labor, Employment and Training Administration. http://www.urban.org/UploadedPDF/411907_registered_apprenticeship.pdf. [September 16, 2017].

Muehlemann, Samuel, and Stefan C. Wolter. 2014. "Return on Investment of Apprenticeship Systems for Enterprises: Evidence from Cost-Benefit Analyses." IZA Journal of Labor Policy. http://www.izajolp.com/content/3/1/25. [March 13, 2015].

OECD (Organization for Economic Development and Cooperation). 2009. *Learning for Jobs*. Paris: Author.

OECD. 2010. *Off to a Good Start: Jobs for Youth*. Paris: Author.

Osterman, Paul and Andrew Weaver. 2014. "Skills and Skill Gaps in Manufacturing" In Richard Locke and Rachel Wellhausen, eds. *Production In the Innovation Economy*. Cambridge, MA: MIT Press

Washington State Workforce Training and Education Coordinating Board. 2014. "2014 Workforce Training Results by Program." Olympia, WA. http://wtb.wa.gov/WorkforceTrainingResults.asp. [March 12, 2016].

XII. LERA Annual Reports

LERA Executive Board Meeting Minutes

Sunday, January 8, 2018, 10:15 a.m.
Marriott Philadelphia Downtown, Philadelphia, PA

LERA officers in attendance at the meeting were: Harry Katz (President), Janice Bellace (Past President), Bill Canak (NCAC Chair), Ariel Avgar (Editor-in-Chief).

Board Members in attendance were: Annette Bernhardt (called in) Matthew Bodah, John Budd (RVP Mid), Bill Dirksen, Eric Fidoten, Michelle Hoyman (RVP East), Joan Husted (called in), Brad Markell, Dan Marschall, Sheila Mayberry, Jim Pruitt (RVP West), Saul Rubinstein, Jeff Wheeler.

Committee Chairs in attendance were: Marlene Heyser (called in), Development Committee Co-Chair and Jeannette Wicks-Lim, LERA@ASSA Meeting Program Committee Co-Chair. LERA staff Emily Smith and Bernadette Tiemann also attended.

Board members who were unable to attend were: Sylvia Allegretto, Virginia Doellgast, Jonathan Donehower, James Hayton, Charles Jeszeck and Patrice Mareschal.

Officers unable to attend were Craig Olson (Secretary-Treasurer) and Kris Rondeau (President Elect).

Call to order. The meeting was called to order at 10:20 am ET by Harry Katz, LERA President.

Approval of the Minutes

LERA President Harry Katz motioned to approve the minutes of the June 2, 2017 LERA Executive Board Meeting that took place in Anaheim, CA. Janice Bellace seconded the motion and the minutes were unanimously approved by the board.

President's Report

Harry Katz, LERA President, reported that he has been involved with LERA for several decades and ILERA as well. Several years ago, LERA had concerns about our finances with deep deficits and questions about the sustainability of our business model. The feeling now at LERA, is that we are in good shape financially. Our membership numbers have revived to expected levels, our meetings have been profitable, the quality of our meeting programs have been consistently high, and administration of the organization is on track.

The last five LERA Annual Meetings have gone well, the programs have really taken shape. In putting together this year's program Katz was struck by the quality and strength of the submitted proposals. Many of the most important themes were covered in the submissions. In addition, the LERA@ASSA meeting has had good attendance this year as it did last year.

LERA's challenge in the face of this solid year-end position is to not become complacent; we need to use this as a spring board for continuous improvement and increasing membership. Work is being done to implement the bylaws revisions that occurred in 2016 and in connecting LERA chapters to national. Some of the areas to improve are increasing the numbers of national members who are members of their local chapter, and receiving lists of local chapter members for increased communication at all levels. The Regional Vice Presidents have been elected and have spent the last year structuring their new positions and organizing their regions. LERA has received good feedback from most local LERA chapters with one major exception, that of Gateway LERA, a topic that we will return to later in the meeting. Katz commented that the success of LERA has to be attributed to the hard work of the board members, officers, committee chairs and other volunteers.

Committee Reports

Financial and Membership Reports—reported by Emily Smith on behalf of Craig Olson who was unable to attend. The good news is that for two years in a row LERA has shown profit. This can be partially attributed to the generosity of Kaiser Permanente for both their sponsorship of LERA and conducting their meeting with us in Anaheim. Year-end numbers are an estimate at this point, but confidence that we will end the year positively. We also significantly increased ad sales in both our printed and electronic publications. Thirdly, our individual members supported the organization with personal contributions in excess of $10,000 more than was expected and we are grateful for this show of support.

Membership numbers are roughly 100 individual members more than this time last year. This strong increase represents a return to our expected membership more than a true increase in our member numbers. In 2014 we saw a rapid decline in membership tied closely to challenges because of a change in our membership software. The numbers reported in our 2014 membership are lower than realistic, but good data is not available for this year. The board was updated on an administrative disruption that occurred in 2014/15 that has now been resolved.

The structure of organizational memberships were discussed including the option of considering a specialized organizational level for university programs. Packaging various items that university programs currently fund has been discussed by the UCIRHRP and the board in the past and it has been determined that the current system of invoicing university programs separately for member dues and various sponsorships is the preferred method. This allows LERA to be responsive to various programs' financial needs or abilities and allows LERA to invoice members individually (they can then request reimbursement from their university programs).

Jeff Wheeler addressed systematically measuring performance indicators with two goals: to maximize revenue but the same time to also shore up participation (perhaps at odds with maximizing revenue). LERA currently measures performance using membership numbers and transaction over various time periods, tracking event participation and registration, tracking individual contributions, tracking sponsorships, and a number of other metrics in regards to income and expenses.

LERA does not have any recommendation to raise membership dues in 2018, due to a recent increase in 2017 and the current financial stability of LERA. NCAC chair Bill Canak commented that membership is not the same value proposition for all constituencies, so perhaps membership structure and dues level should be adjusted accordingly by group, and benefits could be formalized and communicated as well. President Harry Katz called for interested volunteers to work with him on membership growth in the coming year.

Nominating Committee Report—reported by John Budd committee chair. A slate of thirty names was discussed for the LERA 2018 election and in discussion a few changes were made to the slate. The revised slate was considered by areas of professional category. The board was reminded that the slate of nominees is considered highly confidential and that none of the individuals have yet been contacted about their nomination. The committee took into consideration diversity in the slate and LERA is always working to increase diversity along a number of attributes including age, geographic distribution, industry, perspective, gender and race, etc. After the board approves a slate, nominees will be contacted one by one to determine their interest in standing for election. If they decline, the next alternate is contacted.

The committee recommended nominating Dennis Dabney as president elect; Jim Pruitt brought the motion to the floor and Brad Markell seconded it. The motion unanimously passed. Discussion included overwhelming support for the nomination of Dennis Dabney.

Discussion ensued on each professional category (Academic, Labor, Management, Neutral/Government/Other). Some substitutions were made in the areas of academics, government/neutral/other and management. The professional categories of academic, government/neutral/other and labor passed unanimously. In the area of management, the final slate recommendation passed with one opposed.

LERA staff will begin contacting the slate of nominees in the order determined by the nominating committee and the executive board and proceed with the election that will take place in early 2018.

Program Committee Report—reported by Harry Katz, LERA President. LERA received strong, healthy proposals submissions and utilized a new electronic ranking tool for LERA 70th Annual Meeting in Baltimore, Md. The program committee strictly enforced the one appearance per role rule and further restricted participants to only two appearances on the LERA program. Staff is working with session organizers to find replacements as necessary. Restricting participation in this way allows us to invite more individuals to the program and be more inclusive.

Other matters related to the meeting included the following. The program committee has decided to schedule committee meetings on the afternoon of the pre-conference day, Wednesday, June 13, 2018, and ending prior to other events. This will allow committee work to take place unimpeded by concurrent sessions and additionally allow committee members to interact with each other. If a committee chair is unable to make the pre-conference day they can request the committee meeting be held on an alternate date. Other events that will take place include a welcome reception on Thursday evening at the National Aquarium, a pre-game party on Friday evening followed by a baseball game at Camden yards, two luncheons, and the annual labor breakfast. It was suggested that we consider adding a session concerned with sports or specifically the MLB, and to consider reaching out to sporting entities for sponsorships.

No increase in registration fees for the 70th Annual Meeting are proposed. Eric Fidoten suggested adding a "foundation fund" or "support speaker sponsorship fund" that can directly support the meeting that is tax deductible. In addition, he recommended that we call registrants attention to the opportunity to contribute to the "speaker sponsor" fund at the time that they register. Annette suggested adding a non-profit category section for registration (excluding unions, universities and other organizations that typically attend LERA), or alternatively a discount code

The location of the 72nd Annual Meeting in 2020 was discussed. Initially Las Vegas was under consideration, but currently Seattle is the top choice, primarily due to the NW LERA Chapter in Seattle. LERA will reach out to officers of NW LERA to determine their interest and in what ways we can complement their operations if they are in agreement. Then union hotels will be located, if any, and proposals requested. Janice Bellace encouraged communication with attendees to be certain to book their accommodations at the LERA hotel.

Baltimore is the location site for 2018. Cleveland is the location site for 2019, and the location for 2020 is undetermined, but will likely be Seattle. Then John Burton spoke about habitually returning to the DC area due to the speakers and policy makers that are available to attend locally. Saul Rubinstein will send information about a hotel utilized by the AFT that is centrally located and affordable.

LERA@ASSA Program Committee Report—reported by Jeannette Wicks-Lim. She explained that a shift in meeting architecture in 2013 resulted in LERA moving its annual meeting to June, but retaining its 27 sessions in January. However, the ASSA subsequently reduced our session allotment to 18 after our attendance dropped. Since that time, the program committee has worked to increase attendance and the quality of the program and the ASSA has no further plans to reduce our session allotment.

One of the program committee's main challenges is to cater to audience at ASSA meeting while also meeting the needs and interests of LERA members. To meet this challenge a strategy has been adopted to focus on labor economists who will draw an audience reporting on topics of interest to LERA members. The program committee has also started a best paper contest in conjunction with the editorial committee and the ILR review. To broaden our reach we have co-sponsored some sessions with other ASSA organizations and we continue to invite high profile speakers. To further support the meeting, we procure sponsorship funds each year for the welcome reception. In 2018 $4,500 was pledged. The program committee has increased the number of roundtable sessions to promote debate and discussion.

The committee continues to rely on session submissions from the program committee, and would like to encourage publicly submitted session proposals. The chairs of the program committee spend one year as subordinate co-chair, one year as lead co-chair and one year as ex-officio on the committee to provide for continuity of leadership. The make-up of the committee members is structured to adequately represent a diverse constituency. In recent years the committee has had more difficulty in recruiting women candidates. Racial diversity is also a concern.

Last year we rebranded the meeting to "LERA@ASSA" and that change seems to be effective. The meeting is largely mission-driven and is an important program; however, it has not been a profitable meeting in the past. Steps have been taken to match income and expenses of this program area and in 2018 we believe the meeting paid for itself. Two strategies have been employed, the first is to require membership from paper presenters and the second has been to acquire sponsorship funding for the welcome reception.

Editorial Committee Report—reported by Ariel Avgar, Editor-in-Chief. The editorial committee strategy has been to have two research volumes in production at any one time. In 2017 and 2018 we will push up the production cycle to culminate in March, alternating it with the production cycle of POW, which finishes in the fall. We are currently ahead of schedule for the 2018 RV and it should be completed by March. The 2019 issue, edited by Suzanne Bruyere, focuses on disability issues in the workplace. The proposal initially focused on a team of academics that worked together, but they have now included researchers from other organizations and have added a chapter that reviews international policies on disabilities in the workplace. The editorial committee plans to bring proposals to the June meeting to the board to maintain this cycle. If you have suggestions for a research volume theme, please contact Ariel Avgar with your ideas or questions.

The editorial committee has introduced a LERA Best Paper competition in conjunction with the ILR Review. This began in 2016 with paper presented at the LERA Annual Meeting in Minneapolis. The editorial committee reviewed papers from authors who wish to participate and presented their papers at the meeting. In the initial year ten submissions were received, five were sent for review, and two were accepted. In 2017 fourteen submissions were received and it is undetermined how many will be accepted. The ultimate goal is for LERA to have a best papers special volume of the ILR review, but until then we will have a special section within a volume.

The editorial committee continues to promote books in the LER field. At the 2018 annual meeting they have planned a session highlight interesting books in the field; this is a way to both promote LERA authors and attract top minds to the LERA. Additionally, several sessions are planned in 2018 highlighting various portions of the research volume.

Eric Fidoten asked if any of our publication materials are available in languages other than English, to better meet the needs of organizations that operate internationally. In response, Ariel explained that while international and comparative issues are topics written about in LERA publications we do not provide alternate translations at this time.

John Budd noted the lack of diversity in the current volume of the LERA *Perspectives on Work*. To address this concern the editor-in-chief will provide guidance and ensure inclusivity.

Jim Pruitt commented that he appreciates this publication specifically for the opportunity it affords to non-academics to publish. Saul Rubinstein suggested adding a supplement from the publication to our website or even interviews from the authors either written or recorded ("LERA Talks"). Additionally we should be considering other ways to repurpose content for *Perspectives* (session programs, interviews with research volume authors or award winners, etc.). Annette Bernhardt suggested that we think of different ways to distribute the *Perspectives*. The *Perspectives on Work* is currently distributed to members, library subscribers, through direct book sales, and through the Copyright Clearance Center. Additional suggestions are welcome. A follow-up report at the June Meeting addressing additional content and distribution of the POW is recommended.

National Chapter Advisory Council Report—reported by Bill Canak, NCAC Chair. The NCAC brought forth the proposal from Penn State (which runs a substantial online education program) to create a virtual LERA chapter. The risks include cannibalism from current and potential chapter and national members. The upside is that Penn State has expertise in the area of online delivery and is willing to take the project on. The board discussed various aspects of the proposal, and request that the NCAC reaches out to Paul Clark and Penn State to further discuss and refine the proposal (itemize deliverables, methods, pricing, etc.) and that Paul Clark attend the next LERA Board Meeting to discuss the proposal in person. Further, Janice Bellace suggested that we ask the current LERA Chapter Presidents for their input using an online survey, and that if the proposal is carried out, that we require that the LERA website be the delivery system for the online chapter.

LERA NCAC members and RVPs continue to communicate bylaws revisions to chapters. They are finding challenges and realizing that being both patient and proactive is required; changes do not occur overnight. Bill is currently working with the Kentucky chapter to assist them in reestablishing chapter membership and leadership, and also working to connect with the San Francisco chapter.

Soliciting sponsorships from the LERA Chapters for the General Membership Meeting and Awards Ceremony has proven successful; $900 was raised in 2017 at the Anaheim meeting.

Three chapter sessions are planned for the Baltimore meeting, including the Chapter Reps meeting, a chapter administration workshop, and the NCAC meeting. Bill suggested that LERA board members and others review the programs at local chapters; there are many interesting and innovative programs taking place around the country. Chapters have proven a vital piece of the LERA Annual Meeting architecture, providing key assistance in nearly every area of the meeting, which shows the strength and vitality of the local chapters.

Regional Vice Presidents' Report—reported by John Budd, LERA RVP, Mid Region. The RVPs, elected in the last year, are working to structure their new positions and organize their chapters regionally. This starts with contacting chapters, and that has proven challenging, although the three RVPs report mostly positive interactions, with one exception. They are currently engaged in building relationships and seeing how they can be of use to the Chapters. Future aspirations may be to consider whether or not travel funds need to be a consideration for RVPs.

Development Committee—reported by Jim Pruitt and Marlene Heyser, Development Committee Co-Chairs. The Development Committee continues to recommend that each board member consider making an annual contribution of $1000 if that is within their means. Jim reports that with Dennis Dabney heading towards becoming President Elect that Kaiser Permanente can likely be counted upon to make a sizeable contribution. Jim wishes to reach out to unions that Kaiser Permanente works with to develop additional organizational members. The committee also hopes that Ford and UAW will continue to sponsor the organization. Marlene Heyser reported that when organizations support us financially it signals to the field that LERA is a worthwhile cause, and it important in many facets. The 2017 Annual Fund Drive raised $26,293 and we had budgeted for $14,000. Our members' individual contributions made a large impact to LERA's financial wellness.

Industry Councils/Interest Sections Coordinating Committee Report—reported by Joel Cutcher-Gershenfeld, IC/IS Coordinating Committee Chair. Joel discussed his vision of Industry Councils and Interest Sections in 2025, one in which LERA Industry Councils and Interest Sections lead the conversation and dialog. Currently, the state of Industry Councils and Interest Sections does not match this vision. Right now, members have the ability to affiliate with IC/IS when they join or renew their memberships, and then not much happens after that. We are falling short and have work to do. At the last annual meeting, the IC/IS met, and we request time to meet at the next LERA Annual Meeting, even if it is 30 minutes, preferably when other sessions are not meeting. People have begun to see the vision for IC/IS, but we need to solidify co-chairs where they have lapsed, build the website to make it user friendly (who do you contact, and how do you make a proposal if one has interest?), update working lists of IC/IS to be consistent across all environments, build tools to allow co-chairs and co-conveners to contact their members and talk to each other (e-lists, discussion boards, saved searches in Member Directory by IC/IS), and use their information and contributions for LERA's Perspective on Work magazine.

Saul Rubinstein mentioned that the K-12 Education Industry Council needs to be added to online forms. Sheila Mayberry suggested that industry councils may be a great resource for Perspectives on Work, local chapters, meeting sessions, etc. Dan Marschall additionally suggested that the virtual chapter could be a tool for the industry councils as well. Online resources there can be utilized by industry or interest section and used to re-invigorate these groups.

New Business

Myron Taylor Award Proposal—reported by John Burton, past LERA President. At his presentation last year in Anaheim, John Burton proposed that Myron Taylor was someone who supported unions, and that it could be used as an award for LERA to recognize a member of management (business community) who has made a contribution to the continuation of a viable union movement in the United States. The award may not be given each year, but could be awarded in years where merit is identified, and possibly to more than one individual, who have worked together. Janice Bellace brought the motion forward and Jim Pruitt seconded it; the motion passed unanimously. Additionally, it was recommended to consider a separate labor-management partnership award proposal for the June 2018 board meeting, in addition to determining details in reference to the award committee structure and award nomination procedure, and guidelines for selection.

ILERA Presidential Nomination—reported by Janice Bellace, LERA Past President. Janice announced her intention to nominate Harry Katz to run as the next ILERA American President term 2021-2024. The individual needs the support of a national association, so she is requesting LERA's support. Saul Rubinstein seconded and the motion passed unanimously. Harry also proposed the idea of pairing the LERA Annual Meeting with the ILERA meeting that year either back-to-back or partially overlapping. It was also suggested to coordinate with Latin American organizations and the Canadian LERA. However, it is not expected that the Latin American cohorts will participate, and the Canadian ILERA organizations already conduct regional meetings about three times per year.

LERA 70th Anniversary Cartoon Book—reported by Bill Canak, NCAC Chair. Bill suggested that we create a LERA 70th anniversary cartoon book with the New Yorker. Bill will move forward to prepare a business plan/proposal for consideration to the board. Additionally, he will inquire with The New Yorker about a digital version of the book.

Gateway LERA Chapter Discussion—reported by Harry Katz, LERA President. LERA received a letter after the 2016 annual meeting itemizing their disagreement with the bylaws revisions. Subsequently, in 2017, their board voted and decided to end their relationship with LERA. The board discussed the situation, and it was determined that the LERA President should visit the chapter in person, as well as reach out to several of their officers and members to gather more information, and determine the exact nature of their disagreements and see is they can be accommodated. The LERA Gateway Chapter is an important chapter with a long history of achievement, and it is the wish of the board and our officers that we work to resolve our disagreements and continue our relationship.

Membership Initiative—reported by Harry Katz, LERA President. Harry invites all board members to continue thinking about how we can expand our membership, and he requested volunteers for a subcommittee to expand membership. He is also reviewing notes from the April 2016 board meeting that focused on membership initiatives.

Adjournment—The meeting was adjourned by Harry Katz at 2:31 pm.

LERA Executive Board Meeting Minutes

Friday, June 15, 2018, 8:00 a.m.
Hilton Baltimore, Ruth Room, Baltimore, Maryland

Call to order—the meeting was called to order at 8:05 am by Harry Katz, President. Present at the meeting were officers Harry Katz (President), Kris Rondeau (President Elect), Janice Bellace (Past President), Dennis Dabney (President Elect 2019), and Ariel Avgar (Editor-in-Chief). Board members present were John Amman (appointed 2018 to J. Husted's remaining term), Sylvia Allegretto, Annette Bernhardt, Matthew Bodah, Bill Dirksen, Virginia Doellgast, Jonathan Donehower, Patrice Mareschal, Brad Markell, Dan Marschall, Sheila Mayberry, Saul Rubinstein, Jeff Wheeler. RVP's present were John Budd (Mid Region), Michelle Hoyman (East Region), Jim Pruitt (West Region). The UCIRHRP Chair Adrienne Eaton was present. Committee Chairs present were Bonnie Castrey (NCAC Vice Chair and Media Award Committee Co-Chair), Marlene Heyser (Development Committee Co-Chair). Special Guests in attendance included incoming board members Ezio Borchini, Cyndi Furseth, and David Weil. LERA staff in attendance included Emily Smith, Executive Director, and Bernadette Tiemann, Membership and Marketing Coordinator. Unable to attend were Joan Husted, Charles Jeszeck, Eric Fidoten, William Canak (NCAC Chair) and Craig Olson (Secretary-Treasurer).

Reports

Approval of the Minutes—The motion to approve the minutes passed unanimously from the board meeting in January 2018 and the General Membership Meeting in June 2017; Jim Pruitt motioned and Janice Bellace seconded.

President's Report—Harry Katz reported that John Amman had been appointed to finish out Joan Husted's term on the board. Additionally, Ryan Lamare had been invited to attend the board meeting and fill in for Craig Olson, who was unable to attend for health reasons. Ryan had been appointed in an interim capacity to the role of Secretary-Treasurer, and board discussion will be invited soon.

Gateway LERA Report—Gateway LERA requested to withdraw from the LERA last year. Harry Katz reported that throughout the last year, he had visited a number of chapters. Gateway LERA had voiced concern about the bylaws revisions in 2016, which in one sense motivated Harry Katz's local chapter visits. In meeting with many of our local chapters, President Katz found that no other local chapters had problems with the bylaws changes, and some of them were unaware of the bylaws revisions. The chapter in Houston did have questions, but during Harry Katz's visit, their questions were answered and they had a favourable response to the changes. Unfortunately, LERA's efforts to communicate with the Gateway Chapter have not resulted in a reconciliation. It's not at all clear to us if the view if widely held among the members, or if it's an opinion held by a few. We have attempted to open pathways of communication, but in the absence of communication our recommendation is to allow them to withdraw from LERA and discontinue the use of the LERA name. At some point in the future, if they wish to return, LERA would certainly welcome them back to LERA Chapter status. After discussion, it was determined that LERA Gateway's request is accepted; the chapter will be declared dormant and could be reactivated should they wish to do so. Lastly, Brad Markell suggests that we begin a chapter in the region since LERA does need a presence in the region.

Nominating Committee Report—Harry Katz reported that LERA would like to return to a process in which people who are elected have one year of advance notice before they begin to serve their elected term, especially for President-Elect who must formulate a program committee, select a theme, and begin work to prepare their committee for their cycle of work. To do so, we will time the next election for late summer or early fall 2018. When LERA moved our annual meeting from January to June, we lost a bit of our advance notice, and we'd like to return to that space. The motion for the election cycle change was unanimously approved by the board.

In the last election, the following new board members and officers were elected: Dennis Dabney, (President Elect 2018-19); and board members: David Weil (Academic, 2018-2021); Adrienne Eaton (Academic, 2018-2021); Brad Markell (Labor, 2018-2021); Cyndi Furseth (Management, 2018-2021), and Ezio Borchini (Neutral, 2018-2021).

Appointment of Secretary-Treasurer—Ryan Lamare was excused from the meeting and discussion began regarding his appointment as Secretary-Treasurer. Ryan Lamare is a tenured professor at the School of LER at the University of Illinois at Urbana-Champaign. He has been closely connected to this organization, is a past award winner of LERA, and has served as the poster session chair for several years. It's useful for the Secretary-Treasurer to understand the Illinois system since LERA employees are University of Illinois employees. Ryan Lamare's appointment as Secretary-Treasurer was approved unanimously by the board.

Finance and Membership Report—Emily Smith reported on LERA finances on behalf of Craig Olson who was unable to attend. In 2016 and 2017, LERA had positive balances, and LERA is projecting a positive balance for 2018 as well. Major sustaining sponsors are key to LERA's continued financial stability, as well as member dues and meeting income. Sponsorships for the LERA Annual Meeting doubled from 2017 to 2018; those funds were used to support special events associated with the Baltimore event. The LERA 70th Annual Meeting registration stands at 443, which is an increase of 123 from the LERA 69th Annual Meeting in Anaheim. Current member numbers are increasing, and have been increasing steadily since 2016; our current number is about 1,040. From 2005 to 2011, LERA doubled our member dues. During that time, member numbers decreased by half, and dues revenue stayed flat. No dues increases were recommended for 2018 and it was noted that the National Chapter Advisory Council also recommended no chapter dues increases in 2018. It was discussed that our numbers of LERA student members has been on the rise, which is encouraging but that LERA needs strategies for retaining student members to become regular members. Ideas that were discussed were to give younger members some responsibility in committee roles, and establishing a mentoring program and retention program for first-time members and first-time meeting attendees.

LERA 71st Annual Meeting Program Committee Report—Harry Katz reported that the program committee was impressed by the quantity and quality of session proposals they received, and that the meeting structure has now been institutionalized into a well-functioning program. Next year the LERA Annual Meeting will take place in Cleveland, and the committee has discussed which organizations should be partnered with. The committee will review proposals in the Fall and work with organizers to adjust session proposals as needed. The committee will balance perspectives in the program, union-management-neutral voices, and a number of other diversity characteristics. The Senator from Ohio is expected to speak, as is the Mayor of Cleveland. One potential special event venue may be the Rock n' Roll Hall of Fame. Dennis Dabney, Program Committee Co-Chair, has contacts in the area as he lived there for a few years. The Co-Chairs urged board members to consider who we should reach out to submit proposals, partner with, and sponsor the meeting in Cleveland.

Annette Bernhardt asked if Cleveland could be a place where we could bring labor together. The labor center has been serving to connect members of the labor community, and this meeting could serve as a working space for labor leaders to come together and collaborate on specific questions. Additionally, an affiliated pre-conference event could be established. Northeast Ohio regional organizations could be connected with as suggested by Dan Marschall. The LERA Executive Board has a key role to help in designing the program, and the annual meeting is a vehicle for people to learn about LERA at both the national level and local levels.

Dennis Dabney, Program Committee Co-Chair and LERA President Elect, reported that the meeting theme chosen was "Ahead of the Curve: Challenges Posed by New Technologies and the Workforce", although the theme is broadly interpreted and session proposals in many areas are equally considered by the program Committee.

Editorial Committee Report—Ariel Avgar reported that the LERA editorial committee doesn't currently have a student member, and that will be resolved. The ILR Review/LERA Best Paper Competition accepts submissions from papers presented at LERA meetings. These papers then undergo an initial review by the editorial committee, and then by the ILR Review. In 2016, two papers were ultimately published in the ILR Review (Stephen Silvia, UAW, and Alan Benson, UMN). In 2017, twelve papers were submitted, six made it through the initial review, and three of those are still in play. It's important for LERA to link the meetings and publishing in a premiere academic journal.

At the LERA Annual Meeting in 2018, the editorial committee hosted a session on books published in the field. Next year, the editorial committee may propose a panel of journalists to discuss the relationship between academia and journalism.

The LERA Research Volume always has two volumes in production. The 2018 volume is nearing completion, with a few chapters yet to make their way through the process. The 2019 volume is well on its way, looking at employment disability issues. On Thursday, June 14, 2018, the editorial committee meeting provided a proposal for the 2020 issue of Research Volume: "Reimagining the Governance of Work and Employment" edited by Dionne Pohler that discusses traditional ways of thinking about work, new models of work, and commentary from academics and practitioners on those ideas. Jim Pruitt comments that he approves of the idea that this proposal gives non-academics an outlet to be published in this volume. The board unanimously approved the 2020 Research Volume proposal.

The *Perspectives on Work* magazine has moved to work with Editorial Committee, and this year has a much broader array of contributors than in the past.

Sheila Mayberry reports that NAA used their funds to support research grants of up to $50,000 and they are currently looking for proposals. Those interested should contact Dick Fincher.

National Chapter Advisory Council Report—Bonnie Castrey reported that the three RVPs (Regional Vice Presidents) attended the 2018 NCAC/RVP on Thursday, June 14, 2018. The NCAC and RVPs will work together and the RVP role will strengthen the chapter connections with national, and we have three excellent RVPs. The bylaws changes that were made in 2016 have been fully carried out, and we are seeing those changes coalesce: President Elect is cultivating program; RVPs are elected and working; affiliates are functional.

Penn State has proposed the creation of a virtual chapter. They do not intend to take members from other local LERA Chapters and do intend to allow people to become members who are outside of fifty miles from a local LERA chapter. The NCAC recommends that we accept their proposal to begin a virtual LERA chapter, which would begin targeting students at Penn State, and to revisit the progress, challenges, and successes of the newly proposed virtual chapter model at the board meeting in 2019. At that point, the board will discuss if they will recommend that the Penn State virtual chapter begin inviting members to join who are further than fifty miles away from a local LERA chapter. Some local chapters are concerned that removing a personal face-to-face connection may erode the effectiveness of the local chapter. Others discuss that developing new delivery systems is both complimentary and also necessary to keep up with new technology. The motion to begin the virtual chapter for one year, promoting membership to PSU students and alumni, and having the chapter report on its progress, challenges, and successes next year at the board meeting in Cleveland, Ohio for further deliberation by the LERA Executive Board was seconded by Jim Pruitt. The motion passed unanimously.

Development Committee Report—The Development and Contributions Committee recommended that the Executive Board adopt the motion that, in LERA's Fiscal Year 2018 the Executive Board authorizes the expenditure of the following amounts from the following restricted funds, with the direction that those amounts will be used at the discretion of the Executive Director for LERA operating expenses: from the John T. Dunlop Public Policy Fund: $4,000 and from the Walter and Gladys Gershenfeld Publications Fund: $4,000. This motion was approved unanimously.

At the Development Committee meeting on Thursday, June 14, 2018, chaired by Marlene Heyser and Jim Pruitt, a few new committee members attended including Harry Katz and Tom Kochan. In 2018, a major gift/estate planning fundraiser will be planned. The Cleveland meeting will turn our focus to organizations and companies in Cleveland. Tom Kochan is speaking in Portland, Oregon at the LERA meeting in November and Jim Pruitt could organize a fundraising dinner at which Tom Kochan could speak. Also, Jim Pruitt recommends that board members consider individual contributions to LERA to the degree that individuals are able to make a contribution. Harry Katz reported that we send a letter each Fall that goes to all our members to consider a contribution in additional to their annual member dues (our Annual Fund Drive). The committee is also considering reminding people to consider an estate gift. Janice Bellace suggested that for the 75th Anniversary LERA create a diamond anniversary fund.

Industry Councils and Interest Sections Coordinating Committee Report—Joel Cutcher-Gershenfeld reported that Interest Sections do exist but their profiles need to be raised. They were designed to be a different port of entry into the association. Health care, public sector, and K-12 industry councils are very active. Some sections are revitalizing and two new councils are being discussed (sports and entertainment, and food industry). To organize, the establishment of a charter is encouraged. The Healthcare Industry Council was part of the healthy pre-conference day activity at the current meeting. The Dispute Resolution Interest Section created a meeting at the last minute. If one of these organizations wish to have a meeting, so we need to communicate that to the various IC/IS's. The Dispute Resolution Interest Section met on June 14, 2018 and twenty people attended so there is a need to continue organizing these meetings at the LERA Annual Meeting. At the IC/IS meeting that took place on Thursday, June 14, 2018, there was discussion of:

- sending quarterly notices of papers published, court cases of interest to respective IC/IS members
- multiple IC's and IS's submit session proposals for the LERA Annual Meeting
- Session proposal to organize on True Union Management Partnerships (TRUMPS)
- Looking for grant funding for LERA Industry Councils and Interest Sections
- Industry Councils represent the opportunity for practitioners and academics to come together, and discuss best practices.
- Proposal to increase industry council education offerings to two groups: higher education and K-12

LERA Media Award Committee—Bonnie Castrey motioned to ratify the name change of the LERA Media Award to the Ken May Media Award and this motion was unanimously passed by the LERA Board. Ken May was a very active LERA member, and received the first LERA media award in 2016.

LERA 70th Annual Cartoon Book Report—Harry Katz reported on behalf of Bill Canak, who was unable to attend. This project has not seen results and if you have any ability to reach Conde Nast, please contact Harry Katz.

LERA@ASSA Program Committee—Harry Katz reported on behalf of the program committee co-chairs. The Program in 2019 has been announced and LERA board members and officers are encouraged to attend. There will be a LERA board meeting at that time.

LERA Awards Committee—The 2018 LERA Award Winners to be announced on Saturday, June 16, 2018 were as follows: John T. Dunlop Outstanding Scholar Awards: Alan Benson, University of Minnesota-Twin Cities and Tobias Schulze-Cleven, Rutgers University; LERA Fellows: Trevor Bain, University of Alabama; Bob Bruno, University of Illinois at Urbana-Champaign: Cyndi Furseth, PGE; Dan Marschall, AFL-CIO; Craig Olson, University of Illinois at Urbana-Champaign; Paula Voos, Rutgers University; Jeff Wheeler, US DOL; Lifetime Achievement Award Winners: Marcia Greenbaum, Arbitrator; Morris Kleiner, University of Minnesota; David Lipsky, Cornell University; LERA Founder Award: Gladys W. Gruenberg; Outstanding Practitioner Award: Bill Dwyer, Rutgers University; Ken May Media Award: Steven Greenhouse, New York Times (ret.); The McClatchy Group; Samuel Zagoria, Washington Post (posthumous); James G. Scoville Best International Paper Award: Katy Fox-Hodess, University of California, Berkeley; Susan C. Eaton Outstanding Scholar-Practitioner Award: Janice Fine, Rutgers University; Susan C. Eaton Research Grant Award: Rebecca Givan, Rutgers University; Thomas A. Kochan & Stephen R. Sleigh Best Dissertation

Award: Tingting Zhang, Western New England University (Winner) and Katherine Eva Maich, Pennsylvania State University (Honorable Mention).

New Business

2020 Annual Meeting Location—Harry Katz reported that LERA was targeting an appropriate hotel venue in Seattle, but was unable to do so, and that now we are exploring a location in Portland, Oregon. The Seattle chapter would also be interested in participating in a return to Oregon.

ILERA Presidential Nomination—Janice Bellace and LERA have nominated Harry Katz for ILERA President in 2024. In that role, Harry would arrange the World Congress in 2024. If elected, he would need to select a location. Boston, Chicago, and downtown Los Angeles are being considered, and Harry solicits your input on locations. Something to consider would be to linking LERA with the World Congress at that point. Johanna Weststar is doing their regional CIRA meeting.

#MeToo Code of Conduct—Adrienne Eaton discussed a potential code of conduct for the #MeToo movement, and would like LERA to consider a proposal. She will bring a proposal to the next annual meeting.

Next meeting will take place on Sunday, January 6, 2019 in Atlanta, GA, in conjunction with the LERA@ASSA 2019 Meeting.

Adjournment—The meeting was adjourned at 10:14 a.m. by President Harry Katz.

LERA General Membership Meeting and Awards Ceremony

5:30 p.m., June 16, 2018
Holiday 4-5-C, Hilton Baltimore

Call to order. The meeting was called to order at 5:34 p.m. by Harry Katz, President. The President stated that LERA meets twice each year, once in June and once in January. Harry introduced Kris Rondeau, President Elect and Dennis Dabney, the next President Elect.

Committee Reports

Nominating Committee Report. Harry Katz reported that the board approved a change in the election cycle to adjust back to a cycle in which our board members and officers have more time between when they are elected and when their terms begin. The next election will be held in September in order to do so.

Finance and Membership Report. The status of our organization in terms of membership and finance was given by Harry Katz. Our membership is growing, even further than what was reported at lunch, we now have 1,059 members. Strong attendance at the meeting and a number of other things have resulted in a positive balance. Ryan Lamare will be assuming the role of Secretary-Treasurer. He is based at the University of Illinois at Urbana-Champaign.

Development and Contributions Committee Report. Harry Katz summarized the development activity on behalf of Marlene Heyser and Jim Pruitt. Although we have a positive balance, please still be generous with donations as we would like to be generous and have a pool of resources in order to have a scholarship fund to assist people who could not otherwise attend to come to the annual meeting. We conduct an Annual Fund Drive in order to receive individual contributions. Marlene Heyser works to receive sponsorships funds, such as Kaiser Permanente and UAW/Ford Motor Co. In addition, this year we will begin asking for final gifting and estate planning contributions. Tia Schneider Denenberg suggested setting up a frequent flyer points fund and hotel points in order to help people attend the annual meeting.

Editorial Committee Report. Ariel Avgar, Editor in Chief, congratulates the program committee on a terrific conference. He recognized the editorial committee. Several years ago the editorial committee partnered with the ILR Review to host the ILR Review/LERA Best Papers competition, and this year he announces that two papers made it through the process to be published this year, and three papers are in process for next year, and he encourages meeting presenters to submit their papers to the competition from this year's annual meeting.

The LERA 2018 Research Volume will be coming out this Fall "No One Size Fits All" edited by Janice Fine et al, and there were a few terrifically attended sessions on this topic at this conference. The 2019 conference will be edited by Susanne Bruyere, and the board approved the 2020 proposal "Reimaging the .." to be edited by Dionne Pohler. Please distribute these volumes widely, and if you would like to give a gift to someone, please consider using this, and also please submit suggestions for new volume ideas to the editorial committee.

Program Committee (LERA 71st Annual Meeting) Report. Dennis Dabney is the lead program committee co-chair. The committee member is planning the conference in Cleveland for 2019. Dennis Dabney lived in Cleveland for three years; they are currently looking for venues (hall of fame) and session ideas, activities, speakers, topics and themes. If you would like to submit sessions, papers, and panels, please do so by Novembber 15, 2018.

NCAC Chapter Report and Star Award. Bonnie Castrey, NCAC Vice Chair reports the NCAC conducted three meetings at this conference. She introduced three new Regional Vice Presidents, who will be contacting chapter. These new roles were added a year ago to help represent chapter interests. We are asking local chapters to send us their email addresses so we can provide them with newsletters, etc.

Harry Katz reports that our national bylaws now require that all national members become a local chapter member. For those of you who are not near a local chapter, Penn State is exploring the option of creating a virtual chapter. They will start with their own students, and revisit services to others at large in the future.

2018 Awards Ceremony

LERA Founders Award: Bonnie Castrey awarded Gladys Gruenberg with the LERA Founders Award. In 2016 Lois Gray received this award. Gladys Gruenberg is 98, she was a founder of the Gateway Chapter, and worked for the NLRB in World War II. In addition to the award, Gladys is being designated as a Lifetime Member.

Media Award: This award has been renamed this year after Ken May, much beloved and recently departed LERA member and journalist. These awards are now being given out in his name. The first 2018 award is being given to Steven Greenhouse, retired from the New York Times labor reporter for many years. The next Media Award is being given to The McClatchy Group. Accepting is Franco Ordonez, white house correspondent and contributor to the series written on misclassification and wage theft. The third media award is in honor of Sam Zagoria, who died earlier this year, and the committee reviewed his body of work. Accepting on his behalf is his daughter Marjorie Olds. Sam and his wife were married for 76 years. Danielle Paquette is also accepting on Sam's behalf, who is now the labor reporter with the Washington Post. Sam finally retired at 94 as the oldest labor arbitrator.

Best Dissertation Award Committee. Ian Greer presents the Tom Kochan and Steve Sleigh Best Dissertation Award to Tingting Zhang on occupational licensing. The award carries a $1000 check in addition to the plaque. The honorable mention was awarded to Katherine Eva Maich, accepted by Paul Clark on her behalf.

James Scoville Best International Paper Award. Ian Greer presents the top paper in this category to Katy Fox-Hodess for the paper "Locating the Local and National in the Global".

Susan C. Eaton Research Grant. Rebecca Givan and Chris Brooks were awarded a check for $3,000 this year to fund research on organizing the south.

ILR Review/LERA Best Papers Award. Rose Batt and Ariel Avgar recognized two papers published as ILR Review/LERA Best Papers, and they were: "Are Bonus Pools Driven by Their Incentive Effects" Evidence from Fluctation in Gainsharing Incentive" by Alan Benson and Sima Sajjiadianni, and "The UAW's Attempts to Unionize Volkswagon Chattanooga" by Stephen Silvia. Rose Batt thanks the awardees for coming.

LERA Awards Committee. Co-Chairs Doug Kruse and Tia Denenberg presented the 2018 slate of awards. The John T. Dunlop Outstanding Scholar Award was given to Alan Benson and Daniel Gilbert for national research issues and the winner for international contributions was Tobias Schulze-Cleven for international research. The Outstanding Practitioner Awardee was William Dwyer, of Public Service Enterprise Group and Rutgers University. The Susan C. Eaton Outstanding Scholar-Practitioner Award went to Janice Fine of Rutgers University. Academic Fellows, given for outstanding contributions to the field of labor and employment relations for 2018 are Paula Voos (Rutgers University), Trevor Bain (University of Alabama), and Bob Bruno, University of Illinois at Urbana-Champaign, and Craig Olson, University of Illinois at Urbana-Champaign. Practitioner Fellows are Cyndi Furseth (Portland Gas and Electric), David Marschall (AFL-CIO), and Jeffrey Wheeler (U.S. Department of Labor and George Washington University). The committee urges the members to nominate worthy awardees in the future.

LERA Chapter Star Award. NCAC vice chair, Bonnie Castrey presented the 2018 LERA Chapter Star award to Maryland LERA (Dan Altchek, President and Ezio Borchini, Past President, accepting), RU LERA (7 student officers accepting), Long Island LERA (Beverly Harrison, Past President, accepting), Oregon LERA (Janet Gillman, President, and Jon Donehower, Board Member, accepting), and TERRA (Bill Canak unable to attend).

New and Other Business

The next LERA Annual Meeting will be held June 13-16, 2019 in Cleveland, Ohio, in the Westin Cleveland Downtown Hotel, and we will meet next at the LERA@ASSA meeting in Atlanta, GA on January 4-6, 2019. After asking for any new or other Business items and hearing none, President Harry Katz made the comment that bylaws changes passed in 2016 have been implemented and well-received by all but one local chapter. Harry has visited over 12 local chapters in the last year, and have heard from many chapters via the newly established Regional Vice Presidents. Harry Katz thanked the Executive Board and all present and handed over the gavel to Kris Rondeau, the new LERA President. President Katz adjourned the meeting at 5:43 p.m.

Labor and Employment Relations Association

Champaign, Illinois

Financial Statements

For the Years Ended December 31, 2017 and 2016

Feller & Kuester CPAs LLP
Certified Public Accountants
806 Parkland Court, Suite 1
Champaign, IL 61821
217-351-3192

TABLE OF CONTENTS

Feller & Kuester CPAs LLP

Tax - Audit - Bookkeeping
806 Parkland Court, Suite #1 - Champaign, Illinois 61821
Phone - (217) 351-3192 Fax - (217) 351-4135 Email - neal@fellerkuester.com

INDEPENDENT AUDITOR'S REPORT

To the Board of Directors of
Labor and Employment Relations Association
Champaign, Illinois

We have audited the accompanying financial statements of the Labor and Employment Relations Association (a nonprofit organization) which comprise the statements of financial position as of December 31, 2017 and 2016, and the related statements of activities and cash flows for the years then ended, and the related notes to the financial statements.

Management's Responsibility for the Financial Statements

Management is responsible for the preparation and fair presentation of these financial statements in accordance with accounting principles generally accepted in the United States of America; this includes the design, implementation, and maintenance of internal control relevant to the preparation and fair presentation of financial statements that are free from material misstatement, whether due to fraud or error.

Auditor's Responsibility

Our responsibility is to express an opinion on these financial statements based on our audits. We conducted our audits in accordance with auditing standards generally accepted in the United States of America. Those standards require that we plan and perform the audit to obtain reasonable assurance about whether the financial statements are free from material misstatement.

An audit involves performing procedures to obtain audit evidence about the amounts and disclosures in the financial statements. The procedures selected depend on the auditor's judgment, including the assessment of the risks of material misstatement of the financial statements, whether due to fraud or error. In making those risk assessments, the auditor considers internal control relevant to the entity's preparation and fair presentation of the financial statements in order to design audit procedures that are appropriate in the circumstances, but not for the purposes of expressing an opinion on the effectiveness of the entity's internal control. Accordingly, we express no such opinion. An audit also includes evaluating the appropriateness of accounting policies used and the reasonableness of significant accounting estimates made by management, as well as evaluating the overall presentation of the financial statements.

We believe that the audit evidence we have obtained is sufficient and appropriate to provide a basis for our audit opinion.

Opinion

In our opinion, the financial statements referred to above present fairly, in all material respects, the financial position of the Labor and Employment Relations Association as of December 31, 2017 and 2016, and the changes in its net assets and its cash flows for the years then ended in accordance with the accounting principles generally accepted in the United States of America.

Other Matter

Our audit was conducted for the purpose of forming an opinion on the financial statements as a whole. The schedules of functional expenses on pages 16 and 17 are presented for purposes of additional analysis and are not a required part of the financial statements. Such information is the responsibility of management and was derived from and relates directly to the underlying accounting and other records used to prepare the financial statements. The information has been subjected to the auditing procedures applied in the audit of the financial statements and certain additional procedures, including comparing and reconciling such information directly to the underlying accounting and other records used to prepare the financial statements or to the financial statements themselves, and other additional procedures in accordance with auditing standards generally accepted in the United States of America. In our opinion, the information is fairly stated in all material respects in relation to the financial statements as a whole.

Feller & Kuester CPAs LLP

Feller & Kuester CPAs LLP
Champaign, Illinois

August 3, 2018

LABOR AND EMPLOYMENT RELATIONS ASSOCIATION
STATEMENTS OF FINANCIAL POSITION
DECEMBER 31, 2017 AND 2016

	2017	2016
ASSETS		
Current Assets		
Cash and Cash Equivalents	$ 170,028	$ 167,732
Investments	24,709	23,796
Accounts Receivable, Net	-	-
Prepaid Expenses	8,131	10,972
Inventory	5,445	6,622
Accrued Interest and Royalties	7,764	-
Total Current Assets	216,077	209,122
Property and Equipment, Net	2,047	276
TOTAL ASSETS	$ 218,124	$ 209,398
LIABILITIES AND NET ASSETS		
Current Liabilities		
Accounts Payable	$ 583	$ 15,421
Funds Held for Other Organizations	10,411	6,605
Accrued Liabilities	25,170	20,424
Dues Collected in Advance	68,538	61,413
Subscriptions Collected in Advance	10,091	13,584
Deferred Chapter and Organization Dues	9,538	7,800
Other Deferred Revenue	8,000	1,500
Total Current Liabilities	132,331	126,747
Net Assets		
Temporarily Restricted		
Susan C. Eaton Scholar-Practitioner Memorial Fund	29,779	32,413
Kochan-Sleigh Best Dissertation Award Fund	40,899	41,941
Jim Scoville Award Fund	5,147	5,251
John T. Dunlop Public Policy Fund	18,828	20,657
Gladys and Water Gershenfeld Publication Fund	20,997	24,550
Total Temporarily Restricted	115,650	124,812
Unrestricted Net Assets	(29,857)	(42,161)
Total Net Assets	85,793	82,651
TOTAL LIABILITIES AND NET ASSETS	$ 218,124	$ 209,398

See Accompanying Notes.

LABOR AND EMPLOYMENT RELATIONS ASSOCIATION
STATEMENTS OF ACTIVITIES
FOR THE YEARS ENDED DECEMBER 31, 2017 AND 2016

	2017			2016		
	Unrestricted	Temporarily Restricted	Total	Unrestricted	Temporarily Restricted	Total
Support and Revenues						
Membership Dues	$ 133,391	$ -	$ 133,391	$ 115,654	$ -	$ 115,654
Meeting Income	146,441	-	146,441	140,661	-	140,661
Organization Dues & Sponsorships	28,700	-	28,700	40,000	-	40,000
Subscriptions	20,651	-	20,651	18,035	-	18,035
Chapter Fees	7,700	-	7,700	9,750	-	9,750
Member Contributions	23,586	1,899	25,485	13,040	1,570	14,610
Royalties	9,485	-	9,485	4,951	-	4,951
Ad Income	10,100	-	10,100	4,013	-	4,013
Publications	1,095	-	1,095	446	-	446
Mailing List Rental	957	-	957	434	-	434
Administrative Fees	256	-	256	174	-	174
Investment Income	-	913	913	-	651	651
Interest Income	157	427	584	140	323	463
Net Assets Released from Restrictions	12,401	(12,401)	-	2,424	(2,424)	-
Total Support and Revenues	394,920	(9,162)	385,758	349,722	120	349,842
Expenses						
Program Services						
General	178,539	-	178,539	156,097	-	156,097
Meetings	101,036	-	101,036	92,726	-	92,726
Publications	36,811	-	36,811	31,297	-	31,297
Supporting Services						
Management and General	57,029	-	57,029	51,659	-	51,659
Membership Development	9,201	-	9,201	14,758	-	14,758
Total Expenses	382,616	-	382,616	346,537	-	346,537
Change in Net Assets	12,304	(9,162)	3,142	3,185	120	3,305
Net Assets, Beginning of Year	(42,161)	124,812	82,651	(45,346)	124,692	79,346
Net Assets, End of Year	$ (29,857)	$ 115,650	$ 85,793	$ (42,161)	$ 124,812	$ 82,651

See Accompanying Notes.

LABOR AND EMPLOYMENT RELATIONS ASSOCIATION
STATEMENTS OF CASH FLOWS
FOR THE YEARS ENDED DECEMBER 31, 2017 AND 2016

	2017	2016
CASH FLOWS FROM OPERATING ACTIVITIES		
Change in Net Assets	$ 3,142	$ 3,305
Adjustments to Reconcile Change in Net Assets to Net Cash Provided by (Used in) Operating Activities		
Depreciation Expense	539	572
Net Unrealized (Gains) Losses on Investments	(252)	62
(Increase) Decrease in Operating Assets		
Accounts Receivable, Net	-	911
Prepaid Expenses	2,841	(8,676)
Inventory	1,177	107
Accrued Interest and Royalties	(7,764)	2,617
Increase (Decrease) in Operating Liabilities		
Accounts Payable	(14,838)	14,846
Funds Held for Other Organizations	3,806	800
Accrued Liabilities	4,746	4,867
Dues Collected in Advance	7,125	6,671
Subscriptions Collected in Advance	(3,493)	3,100
Deferred Chapter Dues	1,738	1,500
Other Deferred Revenue	6,500	1,500
Net Cash Provided by (Used in) Operating Activities	5,267	32,182
CASH FLOWS FROM INVESTMENT ACTIVITIES		
Reinvested Interest, Dividends, and Capital Gains	(661)	(713)
Purchases of Property and Equipment	(2,310)	-
Net Cash Provided by (Used in) Investment Activities	(2,971)	(713)
CASH FLOWS FROM FINANCING ACTIVITIES		
None	-	-
Net Cash Provided by (Used in) Financing Activities	-	-
Net Increase (Decrease) in Cash and Cash Equivalents	2,296	31,469
Beginning Cash and Cash Equivalents	167,732	136,263
Ending Cash and Cash Equivalents	$ 170,028	$ 167,732

See Accompanying Notes.

NOTE 1 – NATURE OF ACTIVITIES AND SIGNIFICANT ACCOUNTING POLICIES

Organization and Nature of Activities

The Labor and Employment Relations Association (the Organization) was founded in 1947 to encourage research in all aspects of the field of labor, employment, and the workplace. It is a nonprofit scholarly association of academic, labor, business, and neutral communities committed to the full discussion and exchange of ideas between and among its broad constituencies through meetings, publications, and its various electronic listservs and websites. The Organization's National Office is located in Champaign, Illinois and serves the association by planning conferences and meetings and publishing the various researches of its members. The main sources of support and revenues for the Organization are contributions, membership dues and subscriptions, and meeting income.

Subsequent Events

Subsequent events have been evaluated through August 3, 2018, which is the date the financial statements were available to be issued.

Accrual Basis of Reporting

The Organization has chosen to report on the accrual basis of accounting. Accordingly, revenue is recognized when earned and expenses are recognized when incurred in conformity with accounting principles generally accepted in the United States of America (GAAP).

Estimates

The preparation of financial statements in conformity with GAAP requires management to make estimates and assumptions that affect certain reported amounts and disclosures. Accordingly, actual results could differ from those estimates.

Cash and Cash Equivalents

For purposes of reporting cash flows, cash and cash equivalents include all cash and highly liquid investments acquired with an original maturity date of three months or less. Since the penalties of converting certificate of deposits to cash is insignificant, all certificate of deposits have been included with cash and cash equivalents. As of December 31, 2017 and 2016, $90,941 and $101,016 of cash and cash equivalents are temporarily restricted and are subject to donor-imposed stipulations.

Investments

Investments consist of intermediate term bond index funds. All investments are held as temporarily restricted and are subject to donor-imposed stipulations.

Accounts Receivable

Accounts receivable are recorded primarily for outstanding invoices for membership dues and subscriptions. An allowance for doubtful accounts is based on an analysis of expected collection rates determined from experience. The Organization had an allowance for doubtful accounts of $0 and $100 at December 31, 2017 and 2016, respectively.

Inventory

The Organization's inventory of directories, research volumes, proceedings, and perspective magazines is carried at the lower of cost and market value. Cost is determined on the basis of first in – first out.

Property and Equipment

Property and equipment expenditures in excess of $500 are capitalized at cost. Donated property and equipment are capitalized at estimated cost or fair market value at the time of donation. Depreciation of the assets is computed using the straight-line method over their estimated useful lives. The range of estimated useful lives by type of asset is as follows:

Furniture and Equipment	5-7 years

Net Assets

Net assets of the Organization and changes therein are classified and reported as follows:

Unrestricted Net Assets – Net assets that are not subject to donor-imposed stipulations.

Temporarily Restricted Net Assets – Net assets subject to donor-imposed stipulations that may or will be met, either by actions of the Organization and/or passage of time. When a restriction expires, temporarily restricted net assets are reclassified from restrictions. Contributions received with temporarily restrictions that are met in the same reporting period are reported as unrestricted support and increase unrestricted net assets. The Organization had $115,650 and $124,812 of temporarily restricted net assets as of December 31, 2017 and 2016, respectively.

Permanently Restricted Net Assets – Net assets subject to donor-imposed stipulations that they be maintained permanently by the Organization. Generally, the donors of these assets permit the Organization to use all or part of the income earned on any related investments for general or specific purposes. The Organization had no permanently restricted net assets as of December 31, 2017 and 2016.

Grant Revenue

Grant funds received under contracts that require provision of services by the Organization are recognized as revenue to the extent that expenses have been incurred for the purpose specified by the grantor. The excess of grant monies received over expenses incurred is accounted for as deferred revenue. The excess of grant expenditures over grant monies received is accounted for as grant revenue receivable to the extent that additional grant revenue has been awarded and will be paid.

Grant funds received with only general restrictions and public donations are considered to be available for unrestricted use unless specifically restricted by the donor. When such restrictions exist, the revenue is recorded as temporarily restricted until the restrictions are met.

Contributed Services and Goods

Contributed services are reported as contribution revenue and as assets or expenses only if the services create or enhance a non-financial asset (for example property and equipment) or:

- Would typically need to be purchased by the Organization if the services had not been provided by contribution.
- Require specialized skills.
- Are provided by individuals with those skills (such as accounting, financial, construction, educational, electrical, legal, medical, and other services provided by accountants, investments advisers, contractors, teachers, electricians, lawyers, doctors, and other professional and craftspeople).

For the years ended December 31, 2017 and 2016, the value of contributed services meeting the defined requirements for recognition in the financial statements as outlined above were not material and have not been recorded on the financial statements. The Organization does receive free office space from the University of Illinois. However, the value for the use of this office space has not been reported on the financial statements.

Income Tax Status

The Organization is a nonprofit association that is exempt from federal income tax under Section 501(c)(3) of the Internal Revenue Code. In addition, the Internal Revenue Service has determined that the Organization is not a private foundation as defined in Section 509(a)(1) and Section 170(b)(1)(A)(vi) of the Code.

The Organization has evaluated its exposure resulting from uncertain income tax position and determined the exposure is not material to the financial statements. In addition, the Organization is not aware of any tax position for which a significant change is reasonably possible within the next 12 months. Therefore, these financial statements do not include a liability for uncertain tax positions. Upon recognition of a liability for an uncertain tax position, the Organization would recognize interest expense and penalties in operating expenses.

The Organization files information tax returns in the U.S. federal jurisdiction and the state of Illinois. Its federal and Illinois information tax returns prior to fiscal year 2014 are closed. The Organization does not have any tax returns currently under examination by either the Internal Revenue Service (IRS) or any U.S. state jurisdiction.

Membership Dues and Advance Subscriptions Collected

Membership dues and subscriptions are assessed and recognized as revenue based on the life of the dues or subscription.

Functional Allocation of Expenses

The costs of providing the various programs and other activities have been summarized on a functional basis in the statement of activities. Accordingly, certain costs have been allocated among the programs and supporting services benefited.

NOTE 2 – ARRANGEMENTS WITH THE UNIVERSITY OF ILLINOIS

The Organization moved its offices to the University of Illinois at the end of 1999. Under an arrangement with the University, the employees of the Organization are employed by the University. The employees' pension and benefits are part of the University's plans. The Organization then reimburses the University monthly for the cost of its employees.

The University of Illinois holds some cash for the Organization. These "claim on cash" balances were $24,352 and $22,650 as of December 31, 2017 and 2016, respectively. These balances have been included with cash and cash equivalents.

NOTE 3 – SERVICING ARRANGEMENTS TO ADMINISTER BUSINESS

On January 3, 2003, the Organization entered into a servicing arrangement to administer the business of the University Council of Industrial Relations and Human Resources Programs (UCIRHRP). These servicing duties include dues notification and collection, annual meeting arrangement and report preparation, and maintaining a data base and network communications for this separate organization. A separate financial statement is maintained for this company, but UCIRHRP's cash is maintained in the Organization's general bank account. The amount of cash belonging to UCIRHRP in the general account is included on the Organization's balance sheet as Funds Held for Other Organizations. UCIRHRP's cash balance in the Organization's custody was $7,160 at December 31, 2017 and $6,605 at December 31, 2016. The Organization is allowed to keep 20% of dues collected each year as an administrative fee.

The Organization entered into an agreement to administer the PhD Student Consortium. The Organization's duties include collecting donations and issuing stipends for eligible students. The amount of cash belonging to the PhD Student Consortium is included on the Organization's balance sheet as Funds Held for Other Organizations. PhD Student Consortium's cash balance in the Organization's custody was $3,251 at December 31, 2017 and $0 at December 31, 2016.

NOTE 4 – CONCENTRATIONS OF CREDIT RISK

Financial instruments that potentially subject the Organization to credit risk consist principally of checking accounts, money markets accounts, and certificates of deposits at financial institutions. However, management continuously monitors the Organization's balances at financial institutions. Deposits held at any financial institutions were fully insured by the Federal Deposit Insurance Corporation (FDIC) at December 31, 2017 or 2016. The "claims on cash" held with the University of Illinois as stated in Note 2 is not insured since it is not held in a financial institution.

NOTE 5 – TEMPORARILY RESTRICTED NET ASSETS

Susan C. Eaton Scholar-Practitioner Memorial Fund

The Organization set up a memorial fund in honor of an author of a 1998 "Perspectives on Work" article who died on December 30, 2003. At its June 1, 2004 meeting, the Organization approved the establishment of an annual Susan C. Eaton Scholar-Practitioner award and grant to be paid to one or more qualified scholar researchers in even-numbered years or practitioners in odd-numbered years doing research in the labor and employment relations or related field. The Organization's Executive Board directed that 10% of any gifts received each year are unrestricted and can be used for administrative expenses. As of December 31, 2017 and 2016, these temporarily restricted funds totaled $29,779 and $32,413, respectively.

Kochan-Sleigh Best Dissertation Award Fund

On March 27, 2006, the Organization set up a temporarily restricted fund with contributions from a member and matching contributions from General Electric. The fund is restricted for a minimum of fifteen years and may be used to pay for the $1,000 annual best dissertation award and plaque. After the fifteen-year period, the fund may be continued or the amount remaining in the fund may be released and become available to the Organization for unrestricted purposes. The Organization's Executive Board directed that 10% of any gifts received each year are unrestricted and can be used for administrative expenses. As of December 31, 2017 and 2016, these temporarily restricted funds totaled $40,899 and $41,941, respectively.

Jim Scoville Award Fund

On January 9, 2009, the Organization set up a temporarily restricted fund with a $10,000 contribution from the University of Minnesota's Industrial Relations Center to honor a member and retiring professor. The fund was established to pay an annual award of $500 and a plaque for best paper on international and comparative employment issues. The Organization's Executive Board directed that 10% of any gifts received each year are unrestricted and can be used for administrative expenses. As of December 31, 2017 and 2016, these temporarily restricted funds totaled $5,147 and $5,251, respectively.

John T. Dunlop Public Policy Fund

On May 8, 2010, the Organization established a temporarily restricted fund to subsidize a named John T. Dunlop Public Policy Session at future National Policy Forums and/or at the Organization annual meetings. The Organization's Executive Board directed that 10% of any gifts received each year are unrestricted and can be used for administrative expenses. This fund will be so named for five or ten years, at which time the Organization's Executive Board will review the fund to determine whether to continue or modify the fund. As of December 31, 2017 and 2016, these temporarily restricted funds totaled $18,828 and $20,657, respectively.

Gladys and Walter Gershenfeld Publication Fund

On May 8, 2010, the Organization established a temporarily restricted fund to honor the memory of long-time members and past presidents Walter and Gladys Gershenfeld for the purpose of supporting the Organization's electronic and print publications. The fund will be so named for ten years, at which time the Organization will review the fund's purpose and uses, and whether to continue or modify the fund. The Organization's Executive Board directed that 10% of any gifts received each year are unrestricted and can be used for administrative expenses. As of December 31, 2017 and 2016, these temporarily restricted funds totaled $20,997 and $24,550, respectively.

NOTE 6 – INVESTMENTS

Financial Accounting Standards Board ("FASB") Codification 820, Fair Value Measurements, establishes a framework for measuring fair value. That framework provides a fair value hierarchy that prioritizes the inputs to valuation techniques used to measure fair value. The hierarch gives the highest priority to unadjusted quoted prices in active markets for identical assets or liabilities (Level 1 measurements) and the lowest priority to unobservable inputs (Level 3 measurements). The three levels of the fair value hierarchy under FASB Codification 820 are described below:

> Level 1 – Inputs to the valuation methodology are based on unadjusted quoted prices for identical assets or liabilities in active markets that the Organization has the ability to access.
>
> Level 2 – Inputs to the valuation methodology include quoted prices for similar assets or liabilities in active markets, quoted prices for identical or similar assets or liabilities in inactive markets, inputs other than quoted prices that are observable for the asset or liability, and inputs that are derived principally from or corroborated by observable market data by correlation or other means. If the asset or liability has a specified (contractual) term, the Level 2 input must be observable for substantially the full term of the asset or liability.
>
> Level 3 – Inputs to the valuation methodology are unobservable and significant to the fair value measurement.

The asset's and liability's fair value measurement level within the fair value hierarchy is based on the lowest level of any input that is significant to the fair value measurement. Valuation techniques used need to maximize the use of observable inputs and minimize the use of unobservable inputs.

The following is a description of the valuation methodologies used for assets measured at fair value. There have been no changes in the methodologies used at December 31, 2017 and 2016.

Intermediate Term Bond Funds: Valued at the NAV of shares held by the Organization at year-end based on readily determinable fair values, which are published daily and are the basis for current transactions.

The preceding methods described may produce a fair value calculation that may not be indicative of net realizable value or reflective of future fair values. Furthermore, although the Organization believes its valuation methods are appropriate and consistent with other market participants, the use of different methodologies or assumptions to determine the fair value of certain financial instruments could result in different fair value measurements at the reporting date.

The following table sets forth by level, within the fair value hierarchy, the Organization's assets at fair value as of December 31, 2017 and 2016:

Assets at Fair Value as of December 31, 2017

	Level 1	Level 2	Level 3	Total
Intermediate Term Bond Funds	$24,709	-	-	$24,709

Assets at Fair Value as of December 31, 2016

	Level 1	Level 2	Level 3	Total
Intermediate Term Bond Funds	$23,796	-	-	$23,796

Net investment earnings for the years ended December 31, 2017 and 2016 are summarized as follows:

	2017	2016
Dividends	$ 637	$ 618
Capital Gains Distributions	24	95
Net Unrealized Gains (Losses)	252	(62)
Total Investment Income	$ 913	$ 651

This investment income, including unrealized gains and losses, are being reported on the statement of activity as investment income.

NOTE 7 – PROPERTY AND EQUIPMENT

As of December 31, 2017, Property and Equipment consists of:

Furniture and Equipment	$ 15,262
Less: Accumulated Depreciation	(13,215)
Property and Equipment, Net	$ 2,047

As of December 31, 2016, Property and Equipment consists of:

Furniture and Equipment	$ 12,952
Less: Accumulated Depreciation	(12,676)
Property and Equipment, Net	$ 276

Depreciation expense for the years ended December 31, 2017 and 2016 was $539 and $572, respectively.

NOTE 8 – ACCRUED EXPENSES

As of December 31, 2017, Accrued Expenses consist of:

Accrued Payroll	$ 5,202
Accrued Benefits/Taxes	3,207
Accrued Vacation	16,761
Total	$ 25,170

As of December 31, 2016, Accrued Expenses consist of:

Accrued Payroll	$ 5,318
Accrued Benefits/Taxes	1,653
Accrued Vacation	13,453
Total	$ 20,424

NOTE 9 – COMMITMENTS

On February 26, 2013, the Organization signed a twelve-month contract from March 1, 2013 to February 28, 2014 with a company to provide services to manage the EPRN website and provide content for the Organization's website. The contract states that the Organization will pay this company $2,400 per month for a total of $28,800. The Agreement automatically renews for consecutive twelve-month terms, unless the Agreement is terminated by one or both of the parties as set forth herein. Beginning the second cycle and continuing until termination or renegotiation, the Organization will increase the base rate every year by two percent (2%) over the previous year's base rate. For example: $2,448 per month for twelve months starting March 1, 2014; $2,497 per month for twelve months starting March 1, 2015; $2,546 per month for twelve months starting March 1, 2016; and so on.

On January 18, 2017, the Organization signed a new modified twelve-month contract from March 1, 2017 to February 28, 2018 with the same company to provide services to provide content for the Organization's website. The contract states that the Organization will pay this company $1,300 per month for a total of $15,600. The Agreement automatically renews for consecutive twelve-month terms, unless the Agreement is terminated by one or both of the parties as set forth herein. Beginning the second cycle and continuing until termination or renegotiation, the Organization will increase the base rate every year by two percent (2%) over the previous year's base rate. For example: $1,326 per month for twelve months starting March

1, 2018; $1,353 per month for twelve months starting March 1, 2019; and so on.

The Organization has signed contracts with hotels in Anaheim, Baltimore, and Cleveland for the 2017, 2018, and 2019 Annual Meetings. If these contracts were cancelled at December 31, 2017 and 2016, the Organization would have owed $113,655 and $128,000, respectively, to the Anaheim, Baltimore, and Cleveland hotels.

NOTE 10 – PENSION PLAN AND RETIREMENT CONTRIBUTIONS

The University of Illinois (the University) contributes to the State Universities Retirement System of Illinois (SURS), a cost-sharing multiple-employer defined benefit plan with a special funding situation whereby the State of Illinois (the State) makes substantially all actuarially determined required contributions on behalf of the participating employers. SURS was established July 21, 1941 to provide retirement annuities and other benefits for staff members and employees of state universities, certain affiliated organizations, and certain other state educational and scientific agencies and for survivors, dependents, and other beneficiaries of such employees. SURS is considered a component unit of the State of Illinois' financial reporting entity and is included in the State's financial reports as a pension trust fund. SURS is governed by Section 5/15, Chapter 40 of the *Illinois Compiled Statutes*. SURS issues a publicly available financial report that includes financial statements and required supplementary information. That report may be obtained by accessing the website at www.SURS.org.

Benefits Provided. A traditional benefit plan was established in 1941. Public Act 90-0448 enacted effective January 1, 1998, established an alternative defined benefit program known as the portable benefit package. The traditional and portable plan Tier 1 refers to members that began participation prior to January 1, 2011. Public Act 96-0889 revised the traditional and portable benefit plans for members who begin participation on or after January 1, 2011, and who do not have other eligible Illinois reciprocal system services. The revised plan is referred to as Tier 2. New employees are allowed 6 months after their date of hire to make an irrevocable election. A summary of the benefit provisions can be found in the SURS' comprehensive annual financial report (CAFR) Notes to the Financial Statements.

Contributions. The State of Illinois is primarily responsible for funding SURS on behalf of the individual employers at an actuarially determined amount. Public Act 88-0593 provides a Statutory Funding Plan consisting of two parts: (i) a ramp-up period from 1996 to 2010 and (ii) a period of contributions equal to a level percentage of the payroll of active members of the System to reach 90 percent of the total Actuarial Accrued Liability by the end of fiscal year 2045. Employer contributions from "trust, federal, and other funds" are provided under Section 15-155(b) of the Illinois Pension Code and require employers to pay contributions which are sufficient to cover the accruing normal costs on behalf of applicable employees. The employer normal cost was 12.53 percent of employee payroll during the period of July 1, 2016 to June 30, 2017. The employer normal cost was 12.46 percent of employee payroll during the period of July 1, 2017 to June 30, 2018. The normal cost is equal to the value of current year's pension benefit and does not include any allocation for the past unfunded liability or interest on the unfunded liability. Plan members are required to contribute 8.0 percent of their annual covered salary. The contribution requirements of plan members and employers are established and may be amended by the Illinois General Assembly. During the years ended December 31, 2017, 2016 and 2015, the Organization reimbursed the University of Illinois $14,764, $10,388, and $13,231, respectively, for the employer's payments to SURS for eligible employees.

In addition to providing pension benefits, the State of Illinois provides certain health, dental and life insurance benefits to annuitants. This includes annuitants of the Organization. Substantially all State employees, including the Origination's employees, may become eligible for postemployment benefits if they eventually become annuitants. Health and dental benefits include basic benefits for annuitants under the State's self-insurance plan and insurance contracts currently in force. Life insurance benefits for annuitants under age 60 are equal to their annual salary at the time of retirement; life insurance benefits for annuitants age 60 or older are limited to $5,000 per annuitant. Currently, the State does not segregate payments made to annuitants from those made to current employees for health, dental and life insurance benefits. These costs are funded by the State and are not an obligation of the Organization.

Employees of the Organization may also elect to participate in several tax deferred annuity plans and defined contribution plans. These are single employer plans under which benefits are provided to participating employees through contracts issued to each individual. Participation and the level of employee contributions are voluntary. The Organization is not required to make contributions.

LABOR AND EMPLOYMENT RELATIONS ASSOCIATION
SCHEDULE OF FUNCTIONAL EXPENSES
FOR THE YEAR ENDED DECEMBER 31, 2017

		MEETINGS			PUBLICATIONS			SUPPORTING SERVICES		
	General	Annual Conference	ASSA Meeting	Other Meetings	Perspectives	Research Volume	Other Publications	Management & General	Membership Development	Totals
Compensation	$ 117,626	$ -	$ -	$ -	$ -	$ -	$ -	$ 18,688	$ -	$ 136,314
Payroll Taxes & Fringes	42,821	-	-	-	-	-	-	6,946	-	49,767
Contract Labor	18,092	-	-	-	-	-	-	-	-	18,092
Depreciation	-	-	-	-	-	-	-	539	-	539
Insurance	-	-	-	-	-	-	-	2,617	-	2,617
Bank and Service Charges	-	-	-	-	-	-	-	6,907	-	6,907
Promotion	-	3,623	-	-	-	-	-	-	1,156	4,779
Fulfillment	-	-	-	-	-	-	-	5,238	-	5,238
Postage and Freight	-	1,127	127	-	1,550	1,610	-	916	-	5,330
Accounting/Auditing	-	-	-	-	-	-	-	4,872	-	4,872
Printing and Production	-	2,014	-	-	7,232	6,441	-	-	-	15,687
Services	-	-	-	-	11,733	6,159	1,557	-	-	19,449
Other Publication Costs	-	-	-	-	257	-	272	-	-	529
Meals/Receptions	-	73,302	5,465	-	-	-	-	-	-	78,767
Travel	-	4,110	455	-	-	-	-	-	-	4,565
Lodging	-	4,896	-	-	-	-	-	-	-	4,896
Other Meeting Expenses	-	3,876	471	1,570	-	-	-	-	-	5,917
Computer Supplies/Services	-	-	-	-	-	-	-	4,434	-	4,434
Website	-	-	-	-	-	-	-	4,569	-	4,569
Office Supplies	-	-	-	-	-	-	-	481	-	481
Student & Member Awards	-	-	-	-	-	-	-	-	6,407	6,407
Fundraising Expense	-	-	-	-	-	-	-	-	871	871
Duplicating Expense	-	-	-	-	-	-	-	476	-	476
Telephone & Fax	-	-	-	-	-	-	-	327	-	327
Chapter Expenses	-	-	-	-	-	-	-	-	363	363
Other Committee Expenses	-	-	-	-	-	-	-	-	404	404
Miscellaneous Office	-	-	-	-	-	-	-	19	-	19
	$ 178,539	$ 92,948	$ 6,518	$ 1,570	$ 20,772	$ 14,210	$ 1,829	$ 57,029	$ 9,201	$ 382,616

LABOR AND EMPLOYMENT RELATIONS ASSOCIATION
SCHEDULE OF FUNCTIONAL EXPENSES
FOR THE YEAR ENDED DECEMBER 31, 2016

		MEETINGS			PUBLICATIONS			SUPPORTING SERVICES		
	General	Annual Conference	ASSA Meeting	Other Meetings	Perspectives	Research Volume	Other Publications	Management & General	Membership Development	Totals
Compensation	$ 76,785	$ -	$ -	$ -	$ -	$ -	$ -	$ 17,382	$ -	$ 94,167
Payroll Taxes & Fringes	26,979	-	-	-	-	-	-	7,533	-	34,512
Contract Labor	52,333	-	-	-	-	-	-	-	-	52,333
Depreciation	-	-	-	-	-	-	-	572	-	572
Insurance	-	-	-	-	-	-	-	3,996	-	3,996
Bank and Service Charges	-	-	-	-	-	-	-	6,980	-	6,980
Promotion	-	5,000	-	-	-	-	-	-	5,143	10,143
Fulfillment	-	-	-	-	-	-	-	4,904	-	4,904
Postage and Freight	-	384	42	-	1,516	1,377	-	1,554	-	4,873
Accounting/Auditing	-	-	-	-	-	-	-	4,865	-	4,865
Printing and Production	-	2,204	74	-	5,068	6,854	102	-	-	14,302
Services	-	-	-	-	9,073	7,238	-	-	-	16,311
Other Publication Costs	-	-	-	-	69	-	-	-	-	69
Meals/Receptions	-	51,927	8,329	213	-	-	-	-	-	60,469
Travel	-	2,445	310	967	-	-	-	-	-	3,722
Lodging	-	10,973	-	-	-	-	-	-	-	10,973
Other Meeting Expenses	-	7,486	1,819	553	-	-	-	-	-	9,858
Computer Supplies/Services	-	-	-	-	-	-	-	88	-	88
Website	-	-	-	-	-	-	-	1,281	-	1,281
Office Supplies	-	-	-	-	-	-	-	883	-	883
Student & Member Awards	-	-	-	-	-	-	-	-	4,794	4,794
Fundraising Expense	-	-	-	-	-	-	-	-	2,117	2,117
Duplicating Expense	-	-	-	-	-	-	-	1,150	-	1,150
Telephone & Fax	-	-	-	-	-	-	-	426	-	426
Chapter Expenses	-	-	-	-	-	-	-	-	132	132
Other Committee Expenses	-	-	-	-	-	-	-	-	2,572	2,572
Miscellaneous Office	-	-	-	-	-	-	-	45	-	45
	$ 156,097	$ 80,419	$ 10,574	$ 1,733	$ 15,726	$ 15,469	$ 102	$ 51,659	$ 14,758	$ 346,537

LERA gratefully acknowledges the continuing support of its Sustaining Sponsors and Organizational Members

SUSTAINING SPONSORS 2018

UAW/Ford National Programs
Kaiser Permanente

ANNUAL MEMBERS 2017-2018

American Federation of Teachers, AFL-CIO
Berkeley Research Group, Labor and Employment Practice
Boston University, Questrom School of Business
Brandeis University, Heller School for Social Policy and Managemen
Cornell University, Institute on Conflict Resolution
Cornell University, School of Industrial and Labor Relations
Federal Mediation and Conciliation Service
Florida International University - College of Business
Ford Motor Company
Kaiser Permanente
Major League Baseball
Massachusetts Institute of Technology, Sloan School of Management
Pennsylvania State University, School of Labor and Employment Relations
Rollins College, Department of Business & Social Entrepreneurship
Rutgers School of Management and Labor Relations
Saint Joseph's University, Erivan K. Haub School of Business
Scheinman Arbitration and Mediation Services
Society for Human Resources Management
The Dannon Company
U.S. Department of Agriculture
United Auto Workers International
United Steelworkers of America
University of California-Berkeley, Center for Labor Research and Education
University of California-Berkeley, Institute for Research on Labor and Employment
University of California-Los Angeles, Institute for Research of Labor and Employment
University of Illinois at Urbana-Champaign, School of Labor and Employment Relations
University of Minnesota, Center for Human Resources and Labor Studies
W.E. Upjohn Institute for Employment Research
West Virginia University - College of Business and Economics

For more information about Organizational Membership with the LERA, please visit the LERA website: http://www.lera.illinois.edu/about/OrgMemInfo.html.

ALPHABETICAL LIST OF AUTHORS

You are invited to become a member of

THE LABOR AND EMPLOYMENT RELATIONS ASSOCIATION

The Labor and Employment Relations Association (LERA) was founded in 1947 by a group who felt that the growing field of industrial relations required an association in which professionally minded people from different organizations could meet. It was intended to enable all who were professionally interested in industrial relations to become better acquainted and to keep up to date with the practices and ideas at work in the field. To our knowledge there is no other organization that affords the multiparty exchange of ideas we have experienced over the years–a unique and valuable forum. After 70 years, both our academic and practitioners agree with the conviction of the founders that the encouragement, reporting, and critical discussion of research are essential if our professional field is to advance.

Our local and national membership of 5,000 includes representatives of management, unions, government; practitioners in consulting, arbitration, mediation, and law; and scholars and teachers representing many disciplines in colleges and universities in the United States and Canada, as well as abroad. Libraries and institutions interested in the publications of the Association are also invited to become subscribing members and thereby receive the same member publications. Organizational memberships in the Association are also available.

Membership dues cover publications for 12 months and entitle members to the electronic newsletter, membership directory and proceedings of the annual meeting online, and printed copies of the annual research volume, and the magazine, *Perspectives on Work* (volumes also available online). Additional online resources include Chapter Profiles and an IR/HR Degree Program Listing for the United States, Canada, and Australia.

Dues for the 2019 calendar year are listed below. International memberships require an additional $25.00 per year to cover postage costs.

Regular Member..$199.00
Contributing Member $310.00
Emeritus Member (10 years+ membership and no longer employed) $110.00
Student Member (full-time, limited to 4 consecutive years) $30.00
Library Subscriber..$300.00

Tax-deductible financial contributions to the Association to support its educational activities are always welcome. You can make an online contribution at our website, leraweb.org, and choosing "Donate". If you are not already a member, we invite you to join the LERA by sending your membership application and dues payment. You can also join online at the LERA website, leraweb.org by choosing "Membership", or request an invoice. Inquiries regarding membership, meetings, and publications should be addressed to the LERA office.

LABOR AND EMPLOYMENT RELATIONS ASSOCIATION
University of Illinois at Urbana-Champaign
121 LER Building, 504 E. Armory Avenue
Champaign, IL 61820 USA
Phone: 217-333-0072 Fax: 217-265-5130

Email: LERAoffice@illinois.edu Web: leraweb.org

Also in the
Labor and Employment Relations Association Series

No One Size Fits All: Worker Organization, Policy and Movement in a New Economic Age
Edited by Janice Fine, Linda Burnham, Kate Griffith, Minsun Ji, Victor Narro, Steven Pitts, and Teyo Reyes..........$34.95

The Contradictions of Pension Fund Capitalism
Edited by Kevin Skerrett, Chris Roberts, Johanna Weststar, and Simon Archer $29.95

The Evolving Health Care Landscape: How Employees, Organizations, and Institutions are Adapting and Innovating
Edited by Ariel C. Avgar and Timothy J. Vogus.......... $29.95

Inequality, Uncertainty, and Opportunity: The Varied and Growing Role of Finance in Labor Relations
Edited by Christian E. Weller..........$29.95

The Disunited States of America: Employment Relations Systems in Conflict
Edited by David Jacobs and Peggy Kahn..........$29.95

Collective Bargaining Under Duress: Case Studies of Major U.S. Industries
Edited by Howard Stranger, Ann C. Frost, and Paul F. Clark..........$29.95

Public Jobs and Political Agendas: The Public Sector in an Era of Economic Stress
Edited by Daniel J. B. Mitchell..........$29.95

Employee Ownership and Shared Capitalism: New Direction in Research
Edited by Edward J. Carberry..........$29.95

Transforming the U.S. Workforce Development System: Lessons from Research and Practice
Edited by David Finegold, Mary Gatta, Hal Salzman, and Susan J. Schurman..........$29.95

Human Rights in Labor and Employment Relations: International and Domestic Perspectives
Edited by James A. Gross and Lance Compa..........$29.95

Book orders and request for prices and order forms for the above LERA publications should be sent to:

ILR Press, Cornell Univ., P.O. Box 6525, Ithaca, NY 14851
Phone 607-277-2211, Fax 607-277-6292
Web www.cornellpress.cornell.edu
Additional volumes are available through ILR Press.

Information about ordering printed copies of the *Proceedings of the LERA Annual Meeting* can be found online at https://lera.memberclicks.net/publications. Orders for printed copies of *Perspectives on Work* and other LERA publications should be sent to the LERA Office:

LERA, 121 LER Building, 504 E. Armory Ave., MC-504, Champaign, IL 61820
Phone 217-333-0072, Fax 217-265-5130
Email LERAoffice@illinois.edu, Web leraweb.org